Pharmaceutical
Pre-Approval Inspections

DRUGS AND THE PHARMACEUTICAL SCIENCES
A Series of Textbooks and Monographs

Executive Editor

James Swarbrick
PharmaceuTech, Inc.
Pinehurst, North Carolina

Advisory Board

Pharmaceutical
Pre-Approval Inspections
A Guide to Regulatory Success
Second Edition

edited by

Martin D. Hynes III
Eli Lilly and Company
Indianapolis, Indiana, USA

CRC Press
Taylor & Francis Group
Boca Raton London New York

CRC Press is an imprint of the
Taylor & Francis Group, an **informa** business

First published 2008 by Informa Healthcare, Inc.

Published 2019 by CRC Press
Taylor & Francis Group
6000 Broken Sound Parkway NW, Suite 300
Boca Raton, FL 33487-2742

First issued in paperback 2019

No claim to original U.S. Government works

ISBN 13: 978-0-367-45271-1 (pbk)
ISBN 13: 978-0-8493-9184-2 (hbk)

Visit the Taylor & Francis Web site at
http://www.taylorandfrancis.com

and the CRC Press Web site at
http://www.crcpress.com

Library of Congress Cataloging-in-Publication Data

Pharmaceutical pre-approval inspections : a guide to regulatory success
/ edited by Martin D. Hynes III. — 2nd ed.
    p. ; cm. — (Drugs and the pharmaceutical sciences ; 181)
  Rev. ed. of: Preparing for FDA pre-approval inspections. c1999.
  Includes bibliographical references and index.
  ISBN-13: 978-0-8493-9184-2 (hardcover : alk. paper)
  ISBN-10: 0-8493-9184-9 (hardcover : alk. paper) 1. Drugs—Research—
Standards—United States. 2. United States. Food and Drug
Administration. I. Hynes, Martin D. II. Preparing for FDA pre-approval
inspections. III. Series.
  [DNLM: 1. United States. Food and Drug Administration. 2. Drug
Industry—standards—United States. 3. Drug Approval—United States.
4. Laboratories—standards—United States. W1 DR893B v.181 2008 / QV
736 P53616 2008]
  RM301.27.P74 2008
  615'.19—dc22

                2007042240

Preface

The regulations that govern the development and manufacture of new drugs have evolved and changed since their advent in the early 1900s. In addition to official changes in the written regulations or laws that apply to the development and manufacturing of drugs, there have been changes in the interpretation of these regulations over time. During the past several decades, the interpretation and enforcement of these regulations have undergone dramatic alterations, significantly affecting the conduct of pharmaceutical research and development. Compliance issues continue to be problematic, at best, and expensive, at worst, costing pharmaceutical developers and manufacturers millions of dollars every year. One example where dramatic change in compliance regulation occurred was in the late 1980s in response to the generic drug scandal. The advent of pre-approval inspections as well as the increased focus on detection and prosecution of fraud evidenced this change. These changes had a significant impact on the submission and timely approval of new drugs. The fact that compliance issues can be costly is demonstrated by the FDA issuing consent decrees in concert with significant fines, some in the neighborhood of a half a billion U.S. dollars over the last decade.

This book focuses on preparing for pre-approval inspections, in light of current trends in FDA expectations as well as enforcement activities with the goal of enhancing the chances of regulatory success thereby avoiding costly compliance issues. In particular, it takes into account the FDA's quality systems-based approach to inspections, risk-based inspections, and the GMP's of the 21st century which, when taken in concert, appear to be a significant alteration in the regulations that guide drug development. Various chapters concentrate on preparation for pre-approval inspections at the start of the drug development process as opposed to the final days before an FDA pre-approval inspection, regardless if it is conducted at a domestic or international site.

In general, the book discusses traditional product development and submission activities for new molecular entities. However, where appropriate, specific guidance is provided to assist in approvals following the technology transfer from one site to another. Particular attention is given to those cases where the transfer is from a domestic site to an international site.

The overall goal of this text is to help the reader prepare for an FDA pre-approval inspection so that the inspection will produce a rapid regulatory approval for the timely delivery of new high quality therapies, to patients in need, and not become an impediment.

Martin D. Hynes III

Acknowledgments

The editor would like to extend his most sincere appreciation to all of the authors who made this work possible. Their willingness to take on the task of authoring a chapter for this book is most appreciated in light of their many other industry and academic commitments.

The editor would also like to thank many of the students and scientists at the institutions from which the chapters originated. Their work and contributions, although not directly cited here, contributed to the advancement of our science in its service to patients. This is especially true at Purdue University where special thanks go to Teresa Cadwallader and at Wyeth where Mike Yelvigi made a number of important contributions to this work.

The editor would like to thank Kate Struck for lending her artistic talents to designing the cover art graphic.

A special thanks goes to Jean Buckwalter who not only co-authored two of the chapters but helped immensely in the preparation of this work. Without her help this volume would not have been possible.

I would also like to thank Sherri Niziolek and Sandy Beberman of Informa Healthcare for their help, patience, and understanding in the preparation of this manuscript.

The editor would like to extend a special and heartfelt thanks to his wife Lynn and daughters Amy and Katie for their love and support during the preparation of this book.

Martin D. Hynes III

Contents

Contributors

Jeanette M. Buckwalter   Eli Lilly and Company, Indianapolis, Indiana, U.S.A.

Graham Bunn   GB Consulting, Berwyn, Pennsylvania, U.S.A.

Tammy Chaney Cullen   Eli Lilly and Company, Indianapolis, Indiana, U.S.A.

Parimal Desai   Wyeth Research, Pearl River, New York, U.S.A.

Marie Crabb-Donat   Eli Lilly and Company, Indianapolis, Indiana, U.S.A.

Julianne Eggert   Eli Lilly and Company, Indianapolis, Indiana, U.S.A.

Mahdi Fawzi   Wyeth Research, Pearl River, New York, U.S.A.

Carol Fowler   Eli Lilly and Company, Indianapolis, Indiana, U.S.A.

David Hoffman   Arnall Golden Gregory LLP, Atlanta, Georgia, U.S.A.

Ludwig Huber   Labcompliance, Oberkirch, Germany

Martin D. Hynes III   Eli Lilly and Company, Indianapolis, Indiana, U.S.A.

Ryan McCann   Department of Industrial and Physical Pharmacy, Purdue University, West Lafayette, Indiana, U.S.A.

Carmen Medina   Precision Consultants, Inc., Coronado, California, U.S.A.

Alan Minsk   Arnall Golden Gregory LLP, Atlanta, Georgia, U.S.A.

Ken Morris   Department of Industrial and Physical Pharmacy, Purdue University, West Lafayette, Indiana, U.S.A.

Lisa Ray   Eli Lilly and Company, Indianapolis, Indiana, U.S.A.

Richard Saunders   Wyeth Research, Pearl River, New York, U.S.A.

Ronald F. Tetzlaff   Parexel Consulting, Alpharetta, Georgia, U.S.A.

Elizabeth M. Troll   Otsuka Pharmaceutical Development & Commercialization, Inc., Rockville, Maryland, U.S.A.

The Evolution of the Food and Drug Administration: Pre–New Drug Application Approval Inspection

Martin D. Hynes III and Jeanette M. Buckwalter
Eli Lilly and Company, Indianapolis, Indiana, U.S.A.

A HISTORICAL OVERVIEW: PAST, PRESENT, AND FUTURE

The Food and Drug Administration's (FDA) pre-approval inspections (PAIs) program is an investigation by the agency to review the adequacy and accuracy of the information provided in a regulatory submission, most frequently the New Drug Application (NDA). This program was first implemented in the FDA's mid-Atlantic region in the late 1980s. Formal communication of the program came in 1990 with a publication entitled ''Mid-Atlantic Region Pharmaceutical Inspection Program'' authored by Henry Avallone (1). Shortly thereafter, the FDA issued a formal compliance manual entitled ''The FDA Compliance Program Guidance Manual (CPGM) on Pre-Approval Inspections/Investigations (Program 7346.832)'' (2). The manual, issued in October 1990, outlined a role for both the Center for Drug Evaluation and Research (CDER) and the district offices in the drug approval process, thus adding the new step of a compliance review to the approval process. The application for marketing approval was now to be reviewed for good manufacturing practices (GMPs) compliance as well as the adequacy and accuracy of the data it included. Prior to this time, companies were on the honor system, with the FDA trusting sponsoring companies to submit accurate data in their marketing application. The advent of the generic drug scandal in the late 1980s brought an end to the honor system, which was replaced by the need to audit all of the data submitted to the FDA for adequacy and accuracy prior to approval of the application (3).

In addition to this historical perspective, these inspections can also be understood within the context of the entire product development process and the myriad of compliance regulations that govern it.

The new product development process for a pharmaceutical begins with discovery and ends when, for all intents and purposes, the drug product is no longer made available to patients in the global marketplace. Numerous scientists and technical experts carry out the product development work, which begins with discovery activities. It is a lengthy and expensive process. This development work results in the delivery of two major classes of "products." The first of these products is the drug product or the material itself, while the second product is information about the new drug product that has been generated during the entire drug development process.

The material that is generated during the drug development process is utilized for a variety of purposes, including preclinical studies, within product development, discovery, as well as in the toxicology functions, and for clinical trials that range from phase I safety studies through phase IV post-marketing surveillance. Given the intended uses of this material in preclinical safety studies, the development process must meet the regulatory requirements that are outlined in the Good Laboratory Practices (GLPs) (4,5), which include documenting synthesis methods and maintaining adequate batch and stability records. The material that is produced for use in clinical trials must meet the regulatory requirements that are prescribed in the Good Clinical Practices (GCPs) (6,7) as well as be manufactured in accordance with the GMPs (8). The GCPs require drug accountability and record keeping, and that material used in clinical trials is manufactured under the GMPs. The 1991 Investigational New Drug (IND) guidelines state that "when drug development reaches the stage when the drug products are produced for clinical trials in humans and animals, then compliance with the GMPs is required" (9). More recent guidance will be reviewed in the chapter.

The information product or deliverable of the new product development process includes all of the data and information that is generated. Data and information that is generated during the development process is generally captured in lab notebooks as well as in a variety of electronic systems, and then summarized in a series of technical reports. These technical reports are written throughout the course of the development process beginning at the initial stages of product development. Toward the end of the product development process, these detailed technical documents and study reports are utilized to generate at least three major types of summary technical documents: the Development History Report, the Common Technical Document (CTD)/NDA, and the Product Development to Manufacturing Technical Transfer document(s). Only a subset of the data generated during the entire development process is extracted and

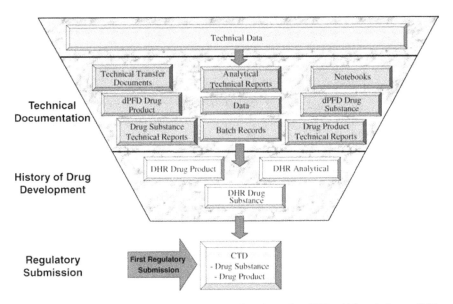

Figure 1   The flow of technical documentation into the CTD. *Abbreviations*: CTD, Common Technical Document; DHR, Development History Report; dPFD, Development Process Flow Document.

utilized in one of these three types of technical reports. A schematic representation of this distillation process from technical report to summary report is depicted in Figure 1. As the information depicted in this figure is generated and summarized during the product development process in technical reports, it must be scientifically sound, well documented, accurate, and compliant with all applicable regulatory requirements which would include the GMPs, GCPs, and GLPs. The quality of this information product is assessed at the end of the development process in at least two important ways, first at the time of technology transfer from the product development organization to the manufacturing component and the other at the time of the regulatory review. This transfer can be judged on the basis of a successful transfer of the product development process and analytical control methods from the product development organization to the manufacturing component. A variety of performance indicators can be used to judge the success of this important transfer; they include ease of product and analytical methods validation, number of failed lots, number of issues or problems encountered in the scale up, the time it takes for the transfer to be successful, as well as the number of issues or complaints that are directed to the product development function.

    The quality of information produced or generated by the product development organization comes at the time of the regulatory review prior to marketing authorization. This regulatory review includes both a scientific assessment

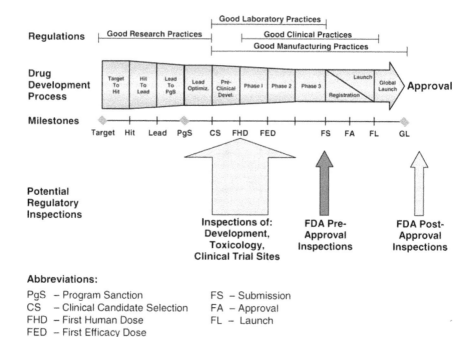

Figure 2   The product development process with points of potential regulatory inspections.

of the information submitted in support of the application as well as a compliance review that occurs during a regulatory inspection like an FDA PAI for marketing authorization. In the United States, the FDA conducts both of these reviews. The scientific review is conducted by CDER, while the compliance review is performed, in large measure by the FDA field compliance groups in concert with other agency groups or functions. (This will be described in greater detail later in the chapter.) These compliance reviews are performed in a general GMP or quality system inspection or through a NDA PAI.

The interrelationship between the drug development process, compliance regulations, and potential regulatory inspections is depicted in Figure 2. As indicated in this figure, it should be noted that in addition to assessments at the time of submission, a compliance assessment of the information or material product from the product development process could literally occur at any time during the entire development process. However for the purpose of this work, we are most concerned with the FDA inspection that occurs at the time of NDA in the United States—the PAI— and hence that will be the focus for the remainder of the chapter. Given this focus, it is important to begin with a definition of what a PAI is.

The PAI is an investigation by one or more FDA investigators to review the adequacy and accuracy of the information provided in the regulatory submission. The program is detailed in the FDA CPGM 7346.832 (10). This program was designed by CDER to cover the following types of drugs and companies:

- New chemical entities
- Drugs within a narrow therapeutic range
- Generic versions of the 200 most prescribed drugs
- Drugs that are difficult to manufacture and replicate
- New dosage form for the application
- First approval for the company
- Poor GMP track record by the company

The object of the program is to:

- Ensure that the facilities listed in the application have the capabilities to fulfill the commitments that have been made to the infrastructure and to process, control, package, label, and test the product.
- Ensure that the manufacturing process is validated.
- Ensure that there is a correlation between the manufacturing process for the material utilized in clinical trials, bioavailability studies, and safety studies with the process that was filed with the submission.
- Ensure that the scientific evidence supports full-scale production procedures and controls.

The FDA's district offices conduct these PAIs. The field investigators from those district offices will audit the data in the submission for both authenticity as well as accuracy. In addition, they will determine the adequacy of the facility, personnel, equipment, and laboratory methods.

Since the program's inception in the late 1980s and early 1990s, there has been an ongoing evolution of the FDA PAI program primarily driven by the issuance of new guidance by the FDA. Some of the major changes that have impacted the PAI program, as well as a number of the changes that may alter the program in future, are depicted on the timeline in Figure 3. The remainder of this chapter will provide an overview of this evolution with the intent of providing a snapshot of the program at this point in time and to discuss the potential for future change.

INTRODUCTION TO THE EVOLUTION OF FDA EXPECTATIONS

There are a number of FDA initiatives currently underway that will significantly alter the development, approval, and manufacturing of new pharmaceutical products. The interconnectivity of all of these changes is depicted in Figure 4. It is not possible within the context of this chapter to provide a detailed review of

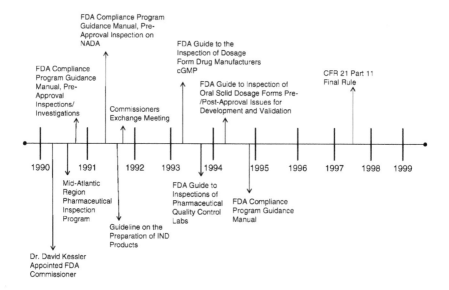

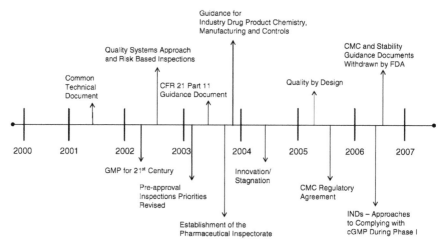

Figure 3 The timeline showing the introduction over the years of the major FDA changes that have impacted the current conduct of PAI. *Abbreviations*: FDA, Food and Drug Administration; PAI, pre-approval inspection; IND, Investigational New Drug; CMC, Chemistry, Manufacturing, and Control.

each of these major initiatives. In fact, there are numerous FDA publications, journal articles, and books that cover each of these topics singly in minute detail. However, we will provide a brief overview of some of the most important of these initiatives to give the reader of this work a high level of understanding of these initiatives and their potential impact to pre-NDA approval inspections in

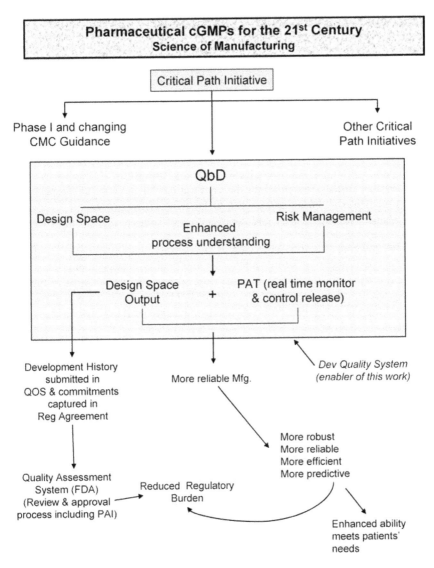

Figure 4 QbD high-level process map. *Abbreviations*: QbD, quality by design; CMC, Chemistry, Manufacturing, and Control; PAT, Process Analytical Technology; QOS, Quality Overall Summary; PAI, Pre-approval inspection.

the future. The initiatives highlighted here are those that will potentially alter product development and the regulatory review process if implemented, and, therefore, are those that are most likely to impact future PAIs to the greatest extent; hence, this is not meant to be an extensive or exhaustive review of all current regulatory change.

PHARMACEUTICAL CGMPS FOR THE 21ST CENTURY: A RISK-BASED APPROACH

One of the FDA initiatives currently underway that will significantly impact the PAI program is that of "Pharmaceutical cGMPs for the 21st Century: A Risk-Based Approach" (11). This major initiative on the regulation of drug product quality was launched in the summer of 2002, and has the following objectives:

- Ensure that regulatory reviews and inspections are based on state-of-the-art science.
- Encourage adoption of new technologies by industry.
- Integrate quality systems into the preparations for review and inspection.
- Implement risk-based approaches to focus on critical areas.
- Enhance the consistency and coordination of drug quality programs.

The purpose of this initiative is to reward manufacturers for their improvements in manufacturing technology and quality improvement initiatives by giving them more flexibility to make post-approval improvements in manufacturing without requiring prior FDA approval (11). In addition, it will allow the FDA to focus its resources on areas where there is the most risk to the final customer—the patient.

For a current status report on this important FDA initiative, please see (12). A multitude of initiatives was generated by this major program, including:

- 21 CFR Part 11 implementation,
- Innovation within the existing framework,
- cGMP warning letter,
- Dispute resolution,
- Aseptic process guidance revision,
- Scientific workshops,
- Comparability protocols,
- Risk-based approaches to the work planning process, and
- Improving the operation of team biologics.

This chapter will focus on the impact of the following initiatives that have been generated as a result of the cGMPs of the 21st Century initiative:

- Pharmaceutical Inspectorate and product specialists,
- PAI compliance program revised priorities,
- Critical Path Initiative,
- INDs—Approaches to complying with GMPs in phase I,
- Quality by Design (QbD),
- Design space,
- Process Analytical Technologies (PAT),
- Quality Overall Summary (QOS),

- Quality Assessment System for chemistry, manufacturing, and control (CMC) Applications,
- Development Quality System, and
- Withdrawal of CMC and stability guidances.

THE ROLE OF THE PHARMACEUTICAL INSPECTORATE AND PRODUCT SPECIALISTS IN THE CONDUCT OF PAIs

Pharmaceutical Inspectorate

The Pharmaceutical Inspectorate was established by the Office of Regulatory Affairs (ORA) and CDER in August 2003 (11). The PI is a group of highly trained field investigators within the ORA who will focus on conducting inspections at prescription drug manufacturers, including PAIs. Their training includes completing a predetermined curriculum to achieve a Level III Drug Certification (13). They are expected to have a basic understanding of the technologies used in manufacturing drugs and be able to apply this knowledge to inspections. Members of the PI will work with product specialists to help ensure that submission reviews and cGMP inspections are coordinated and synergistic. This close working relationship will also help ensure that consistency in regulatory decision-making is strengthened (14).

Product Specialist

The role of the product specialist has been reevaluated since the advent of the FDA cGMPs for the 21st Century initiative. The term ''specialist'' refers to a research scientist or technical expert in an FDA center who is an expert in the science associated with complex products and technologies. These specialists provide scientific support to Investigators during inspections and investigations.

Center for Biologics Evaluation and Research (CBER) specialists have participated in nearly all PAIs since the inception of the program. CBER cGMP experts provide the leadership role in these inspections, whereas the product specialists participating in the inspection bring their specialized expertise in specific scientific and technical areas to bear (11). In particular, a specialist is responsible for the evaluation of test methods, method validation, and analytical data. In the case of biological products, the specialist is responsible for assessing the purification process for the active pharmaceutical ingredient, product purity, and capacity profiles.

Reviewers from the CDER have participated in a number of PAIs over time. Given the value to the Office of New Drug Chemistry (ONDC) review function, their involvement is expected to continue in the future.

Historically, the roles and responsibilities of the specialist during the PAIs for CDER have not been well defined. This has been borne out by actual inspection experience where the responsibilities have varied on the basis of a number of different factors, such as inspection timing, the lead investigator's preferences, pre-inspection planning, investigator/reviewer synergy, and type of inspection (11). As a result, the FDA established a multidisciplinary working group to evaluate the role of the product specialist in PAIs (11). The results of the working group's review indicated that these product specialists have added value to the regulatory process. Thus, a more formalized product specialist program is expected to have a beneficial impact on the cGMPs for the 21st Century initiative. The working group has made a number of detailed recommendations to promote the product specialist program for all drug centers at the FDA. An ongoing and consistent role for product specialists in all future PAIs is, therefore, anticipated.

SCIENCE OF MANUFACTURING

At the inception of the cGMPs for the 21st Century initiative, a working group was established (8). The primary focus of this group has been to find new ways to use the knowledge generated during pharmaceutical development, scale up, optimization, and production while making regulatory risk-based decisions. This working group is currently in the process of identifying efficient approaches for characterizing and controlling critical manufacturing process parameters. Additionally, they are working on a broad regulatory strategy that ensures that both application reviews and cGMP programs are based upon sound, state-of-the-art scientific and engineering knowledge. The goal of this regulatory strategy is to:

- Encourage manufacturers to develop and implement the latest technologies in the pharmaceutical manufacturing process.
- Enhance the use of product and process development technology throughout the product life cycle.

This focus on the product development process will facilitate risk-based regulatory decision-making, innovation, and the use of risk identification management and control methodologies.

The Manufacturing Science working group is charged with the development of a broad regulatory strategy that ensures that NDA reviews and FDA's cGMP programs are based upon sound, state-of-the-art scientific and engineering knowledge (8). The goal of this regulatory strategy is to encourage manufacturers to develop and implement the latest technologies in pharmaceutical manufacturing processes. This will enhance the use of product and process development technology throughout the life cycle of a product. A focus on process understanding will facilitate risk-based regulatory decisions and innovation. In addition, the use of appropriate risk identification management and control methodologies is enhanced.

Thus, one of the overriding themes of the cGMPs of the 21st Century is a risk-based regulatory approach. Additionally, this initiative also has the following objectives:

- Encourage early adoption of new technological advances.
- Facilitate industry implementation of quality management techniques (e.g., quality systems approach).
- Focus industry and agency resources on critical areas.
- Ensure that regulatory review and inspection policies are based on state-of-the-art pharmaceutical science.
- Enhance the consistency and coordination of the FDA's drug quality regulatory programs.

Taking a risk-based approach for inspections will also allow the FDA to allocate its resources to the areas of highest priority, i.e., those areas that will have the greatest impact on public health. Toward that end, the FDA is in the process of developing a quantitative risk-based site selection model for inspection site selection (15). This FDA model will include the assessment of risk factors, such as: the compliance history of the facility, type of drugs manufactured, the process utilized in manufacture, and the overall level of process understanding (15). In order to make this risk-based approach operational, the FDA will revise all of its field compliance programs to incorporate a risk-based approach. One of the programs, as well as accompanying compliance manuals, that has been selected for the incorporation of a risk-based approach is the PAI program.

UPDATING THE CPGM 7346.832 PAIs AND INVESTIGATIONS: PART II—IMPLEMENTATION

The implementation portion of the Compliance Program Guidance Manual for Pre-Approval Inspections (CPGM 7346.832) was updated in September of 2003 to incorporate a risk-based approach to the inspection categories that would prompt a PAI (10). This change is consistent with the agency's cGMPs for the 21st Century initiative as previously described. The goal of these changes is to help reduce the number of PAIs required by the FDA so that the FDA's resources can be used more effectively to conduct post-approval or cGMP inspections.

The risk management approach is being applied to the allocation of FDA inspectional resources for PAIs. The FDA plans to conduct PAIs where they are likely to achieve the greatest public health impact. This model is based on risk factors relating to facilities, compliance history, the type of drugs being manufactured, the manufacturing process being utilized, and the level of process understanding.

Two general categories that exist for the assigning of PAIs include (10):

- Category 1) Those that will regularly prompt an inspection from CDER, which are as follows:
  - New molecular entities,
  - Priority new drug applications,
  - First applications filed,
  - For-cause inspection,
  - Original applications, if current cGMP status is unacceptable or greater than two years,
  - Pre-approval supplements (site change, major construction), if cGMP status is unacceptable,
  - Treatment IND inspections, and
  - Information indicates inspection is warranted to protect patients.

- Category 2) Those where the district office deems the need to inspect:
  - All original applications not previously listed, and
  - All pre-approved supplements not previously listed.

These changes in PAI requirements provide for:

- Greater flexibility for the field offices in determining the need for a PAI on the basis of their knowledge of the firm.
- Deletion of inspection categories for narrow therapeutic range drugs.
- Deletion of the inspection categories for the generic versions of the top 200 most-prescribed drugs.

CDER is anticipating additional changes to the PAI program to allow greater use of risk-based decisions in the assignment and conduct of inspections.

QUALITY ASSESSMENT SYSTEM FOR SUBMISSION OF CHEMISTRY, MANUFACTURING, AND CONTROLS INFORMATION IN A NDA

The FDA Office of New Drug Quality Assessment (ONDQA) has proposed the creation of a regulatory agreement between the agency and the sponsoring company (16). This agreement would be to provide governance for the CMC sections of the NDA, with the intent of providing a structure for regulatory flexibility. The FDA would grant this flexibility to those companies that demonstrate a thorough understanding of the manufacturing process. As currently proposed by the agency, the regulatory agreement would contain both binding elements as well as boundaries for changes. The binding elements of the agreement would include critical process parameters or critical quality attributes in addition to the boundaries of a design space within which the sponsor companies could make changes with no filing of manufacturing supplements. It is anticipated that PAIs would focus on ensuring sponsor companies have a thorough understanding of the manufacturing process. Firms have requested that the regulatory agreement approach be clearly defined, as this enables firms to make clear distinctions in their applications as to what portions of the application are informative (i.e., tell the story of the development of the product and process) and those which are intended to be the binding regulatory commitments going forward.

A new guidance is being proposed by the International Conference on Harmonization (ICH), which will define the modern quality system needed to assure quality over the life cycle of the product (17). ICH Q10 Quality Systems will not be a global GMP guideline but will use existing quality system documents such as ISO and the draft FDA guidances to facilitate the realization of the full benefits of ICH Q8 Pharmaceutical Development and ICH Q9 Quality Risk Management. The scope of this proposed guidance will include both drug substances and drug products over the full life cycle of these products. The life cycle would start with product development, go to technology transfer, to PAI, and on to commercial manufacturing. It is anticipated that this ICH document will provide guidance to industry to ensure that systems are in place to guarantee that the correct decisions are made to manage changes, both within and without the design space.

The ONDC in the Office of Pharmaceutical Science at CDER is in the process of establishing a risk-based quality assessment system for the CMC submission (18). This new quality assessment system is designed to focus on critical pharmaceutical quality attributes related to chemistry, formulation, manufacturing process design, and product performance and their relevance to safety and effectiveness. This system is based upon two other guidance documents: the ICH Q8 Pharmaceutical Development (19) and the PAT—A Framework for Innovative Pharmaceutical Development, Manufacturing, and Quality Assessment (20). This new quality assessment system is designed to facilitate coordination and improvement across the entire product life cycle. Additionally, the system is intended to provide regulatory flexibility for the setting of specifications and post-approval changes based on demonstrated scientific knowledge and understanding of the product and process by applying QbD principles. The quality assessment system aims to focus on the critical quality attributes of the drug product as they relate to the patient. In this regard, the FDA is not focusing on the individual processes of drug manufacturing, but rather on the attributes of the drug product itself.

This new quality assessment system will require that applicants submit in the CMC section of the NDA, the following:

- Information that demonstrates their product knowledge and process understanding,
- Comprehensive Quality Overall Summary,
- More in depth demonstration of process understanding in the Pharmaceutical Development section of the CTD, and
- Information on critical quality attributes and their relationship to clinical performance from both a safety and efficacy standpoint.

The goals of asking applicants to submit this type of information are to:

- Provide a level of confidence to the FDA that quality has been built into the system by demonstrating the extent of product and process knowledge,
- Identify potential sources of variability in manufacture of the product, and
- Explore how risk can be mitigated.

Despite the fact that additional information is being asked for in many areas of the CMC submission, there are areas where less information would need to be submitted to the FDA. These areas could include executed batch records, redundant chromatographic data, and tables of stability data. Since this information would not be included in the submission, it could be reviewed as part of the inspectional activities. The most logical place to inspect these materials is during the course of a pre-NDA approval inspection at the sponsor company. Since the pilot of the new quality assessment system was only started in October of 2005, it is premature at the time of writing this chapter to fully understand the impact of this new assessment system on PAIs as well as additional GMP inspection activities. The reader is encouraged to monitor the new pharmaceutical quality assessment system initiative for its potential impact to PAIs as well as to the submission requirements and process.

THE CRITICAL PATH INITIATIVE

The FDA's concern about increasing cost, long cycle times, and decreasing innovation was also clearly spelled out in "Innovation or Stagnation: Challenges and Opportunity in the Critical Path to New Medical Products" (21). In this document, they argue that the growing cost and difficulties of medical product development will lead to stagnation or a decline of innovation, thus hampering the ability of the biomedical revolution to deliver on the promise of better health.

In the FDA white paper authored by Dr. Janet Woodcock (22), she argues that the medical product development process is no longer able to keep pace with basic scientific innovation. The FDA sees that this is, in part, due to the fact that developers are still using the research tools of the last century to evaluate the discoveries of this century. As an example, they cite toxicology and human safety testing where the tools used are decades old (23). In response to this situation, the FDA has recommended a focused effort to modernize the drug development critical path. This effort would be aimed at improving the medical product development process itself so that the critical development path that leads from scientific discovery to the patient could be streamlined and modernized. The goal of this improved methodology could be to efficiently and predictably produce medical products that are safe and effective. Without the investment in new tools and technologies, society as a whole will be frustrated with both the slow pace of new innovation and the poor yields of the traditional medical development process.

The FDA has indicated that they will work with developers of new medical products to identify and resolve problems encountered during the development process. This work will include both the identification and prioritization of the most significant process issues as well as those areas that will yield the most significant improvements. Once these areas have been identified, the FDA has indicated that it will work with the medical development community to identify and resolve critical development problems with the intent of bringing new drugs to the market more expeditiously.

Significant changes to the medical product development process would be expected to have a corresponding impact on PAIs. The exact nature of that impact is difficult to predict, given that the critical path initiative is still in its infancy. If the FDA initiative is successful in significantly altering and modernizing the drug development process, there could be significant changes to the conduct of PAIs. Given the potential for the critical path initiative to alter the drug development process and, hence, PAIs, the reader is encouraged to keep abreast of this important FDA initiative.

INDs—APPROACHES TO COMPLYING WITH GMPs IN PHASE I

The FDA took a significant step forward at the outset of 2006 by easing the cGMP requirement for the manufacturing of Phase I clinical trial materials. As a result of this action, Phase I drug manufacture was exempted from the requirements of 21 CFR 2.1 "Current Good Manufacturing Practices for Finished Pharmaceuticals" (24). According to Dr. Janet Woodcock, assistant FDA commissioner, "This action is intended to streamline and promote the drug development process while ensuring the safety and quality of the earliest stage investigational drug products" (25).

This rule change was announced shortly after the publication of new draft guidelines on "INDs—Approaches to Complying with cGMPs in Phase I" (26). This document clarifies the FDA's intentions that an incremental approach to control be taken during the development process, thus representing a significant shift from the 1991 Guidance on the Preparation of Investigational New Drug Products (Human and Animal) (27), which required adherence to commercial-scale cGMP standards during the drug development process beginning with Phase I material.

A key question is why the FDA would take such an action at this point in time. It has become clear in recent years that the FDA has become more concerned about the increasing cost and time it takes to bring new drugs to market to meet patient needs as indicated by the discussion of the critical path initiative. This concern is substantiated by the outcome of impact analyses of these changes to the Phase I cGMPs, which are required in advance of implementation. Impact analysis of this rule change conducted by the FDA suggests that the industry could save significant time as a result of the proposed change. The analysis suggested that the industry spends about 848,625 work-hours per year collecting manufacturing data to comply with CFR parts 210 and 211, which will be reduced by 50,000 hours as a result of this rule change (28). Additionally, this concern is further substantiated by the statistics from Dr. Janet Woodcock, FDA deputy commissioner for operations. ''These requirements are so burdensome for early Phase I studies that many leading medical research institutions have not been able to conduct these studies of discoveries made in their laboratories. Today, for the first time, these medical researchers are getting specific advice from the FDA about how to safely prepare products for exploratory studies'' (23).

QUALITY BY DESIGN

Companies with well-designed and implemented quality systems should be well prepared to withstand the scrutiny of a PIA as well as to consistently produce safe and effective products. ''The mutual goal of industry, society, and the FDA is to achieve a desired state of a maximally efficient, agile, flexible pharmaceutical manufacturing sector that reliably produces high-quality drug products without extensive regulatory oversight'' (22). Within this state, manufacturers gain extensive knowledge about critical product and process parameters in addition to quality attributes while striving for continuous improvement. Product quality and performance are assured by effective and efficient manufacturing processes. Product attributes are based on the mechanistic understanding of how the product formulation and manufacturing process impact performance.

The vision for the FDA's role in the future is one of initial verification and subsequent audit with no manufacturing supplements needed. The regulatory policies would recognize the industry's level of product and process knowledge and understanding, which promote continuous improvement. Continuous improvement involves many facets of manufacturing, including:

- Integration of technical innovation and product quality indicators,
- Process understanding and control,
- Variability reduction,
- Quality planning,
- Risk management,
- Corrective action/preventative action, and
- Change management.

There are three key concepts to achieving this desired state: QbD, the design space concept, and the quality system approach.

QbD is a systematic process of building desirable quality attributes into a product to assure its performance (29). These attributes ensure the delivery of drug product with the intended quality and performance. They also assure the identity, purity, quality, and strength/potency of the drug product as it relates to the safety and efficacy of the product throughout the product life cycle. The intent behind QbD is to ensure that industry has a deeper knowledge-based understanding of their drug production processes. QbD requires that the industry develop a scientific understanding of the critical processes and product attributes, design controls, and testing based on that understanding at the time of product development. In addition, industry needs to utilize the knowledge gained over the product life cycle to operate in an environment of continuous improvement, where the manufacturing process is continually being modified to improve production based upon the information gained.

In support of QbD, the FDA would assess the level of product and process understanding to determine if there is sufficient information and knowledge to support the design space. This FDA assessment would initially be done at the time of regulatory review and/or FDA pre-NDA approval inspection.

DESIGN SPACE

The aim of pharmaceutical development is to design a quality product and the manufacturing processes needed to deliver that quality product in a reproducible manner. Control of critical process parameters and material attributes for the production of drug products are paramount to achieving this end. The design space is the established range of process parameters that has been demonstrated to provide assurance of product quality (19). At a minimum, those aspects of drug substances, excipients, and manufacturing processes that are critical and present a significant risk to product quality should be monitored or controlled.

Design space is the multidimensional combination and interaction of input variables (e.g., material attributes) and process parameters that have been demonstrated to provide assurance of quality. The design space is described by the important information on how the product is made and what key attributes are needed to ensure product quality. For example, the design space is concerned with temperature control at a key step in the manufacturing process rather than the type of equipment used for this temperature control.

Changes in formulation and manufacturing processes during development must be seen as opportunities to gain additional knowledge and to support further establishment of the design space. Conducting additional pharmaceutical development studies will lead to an enhanced knowledge of product performance over a wider range of material attributes, processing options, and process parameters. It is this higher degree of understanding of manufacturing processes and controls that will give the FDA confidence to grant additional manufacturing flexibility than would otherwise be available via a traditional application. Opportunities exist to develop more flexible regulatory approaches to facilitate (19):

- Risk-based regulatory decisions (reviews and inspections);
- Manufacturing process improvements, within the approved design space described in the dossier, without further regulatory review; and
- "Real time" quality control, leading to a reduction in end-product release testing.

The level of product and process knowledge that is required at the time of the PAI is demonstrated by a manufacturer's ability to establish a design space for their particular drug product. Future changes in process parameters and formulation attributes will not require additional FDA inspections as long as manufacturers are working within the design space. However, any movement outside of the design space is considered a change that would, under normal circumstances, initiate a regulatory post-approval change process. Thus, innovations are now served by conducting the appropriate studies during the development period to establish a design space that will support future manufacturing activities. These development studies and supporting data will be subject to FDA scrutiny during the PAI.

PROCESS ANALYTICAL TECHNOLOGY

PAT is a system for designing, analyzing, and controlling manufacturing through timely measurements of critical quality and performance attributes of raw and in-process materials and processes with the goal of assuring final product quality (20). These new technologies are for use in both the pharmaceutical manufacturing process as well as in quality assurance activities. The goal of the FDA's PAT approach is to help the pharmaceutical industry enhance its understanding of the manufacturing process. This enhanced knowledge can then be used to meet regulatory requirements for validating and controlling the manufacturing process. The PAT tool is designed to allow the industry to make improvements in manufacturing without fear of regulatory intervention at the smallest change. Implementation of PAT is voluntary, and its implementation for one product does not require it to be implemented for all products.

A desired outcome of the PAT framework is to design and develop well-understood processes that will consistently ensure a predefined quality at the end of the manufacturing process. This guidance facilitates innovation by focusing on process understanding. A process is considered well understood when:

- All critical sources of variability are identified and explained,
- Variability is managed by the process, and
- Product quality attributes can be accurately and reliably predicted over the design space established for materials used, process parameters, manufacturing, environmental, and other conditions (20).

By focusing on process understanding, the burden for validating systems can be reduced by providing more options for justifying and qualifying systems intended to monitor and control attributes of the materials and processes.

There are many tools available that enable process understanding for use in scientific, risk-managed pharmaceutical development, manufacture, and quality assurance. These tools include:

- Multivariate tools for experimental design, data acquisition, and analysis,
- Process analyzers,

- Process control tools, and
- Continuous improvement and knowledge management tools.

The design and optimization of drug formulations and manufacturing processes within the QbD framework can include the following steps:

- Identify and measure critical material and process attributes relating to product quality.
- Design a process measurement system to allow real time or near real time (e.g., on-, in-, or at-line) monitoring of all critical attributes.
- Design process controls that provide adjustments to ensure control of all critical attributes.
- Develop mathematical relationships between product quality attributes and measurements of critical material and process attributes.

Within an established quality system and for a particular manufacturing process, there is an inverse relationship between the level of process understanding and the risk of making a poor quality product. Therefore, by demonstrating a good understanding of the process, there are many opportunities to manage change without requiring regulatory intervention. This is beneficial for both the manufacturer and the FDA.

One potential benefit of utilizing the PAT guideline is real-time release of product. Real-time release is the ability to evaluate and ensure the acceptable quality of in-process and/or final product on the basis of in-process data and information rather than off-line testing after manufacturing is complete. The combined process measurements and other test data gathered during the manufacturing process can serve as the basis for real-time release of the final product and would demonstrate that each batch conforms to established regulatory quality attributes. In fact, process understanding, control strategies, plus on-, in-, or at-line measurement of critical attributes can justify how real-time quality assurance is at least equivalent, if not better than, laboratory-based testing on collected samples. Real-time release, as defined in the PAT Guidance for Industry (20), meets the requirements of testing and release for distribution (30).

One goal of the PAT guidance is to tailor the FDA's usual regulatory scrutiny to meet the needs of PAT-based innovations that:

- Improve the scientific basis for establishing regulatory specifications,
- Promote continuous improvement, and
- Improve manufacturing while maintaining or improving the current level of product quality.

From a technical standpoint, the FDA seeks to assist pharmaceutical manufacturers with the design, development, and implementation of new tools for product manufacture that will allow the pharmaceutical industry to maintain or improve the current level of quality assurance.

PAT along with ICH guidance entitled "Q8 Pharmaceutical Development" are the principles underlying the new quality assessment system for the submission of CMC information in a NDA that is currently being studied by the FDA (31). Reviews of the CMC section of the NDA in concert with the conduct of PAI are the major points of FDA scrutiny that must occur prior to marketing approval. Given the extent to which these two steps are linked, there are likely to be ramifications to PAI given this change to the review of the CMC section of the NDA. The FDA is currently conducting a pilot of the new pharmaceutical quality assessment system with one of the stated objectives being obtaining feedback that will allow the FDA to modify its guidance in this CMC quality assessment system. Following the conduct and assessment of the pilot, it will then become clearer as to the impact of PAT on the conduct of PAIs and on the process of product development itself.

QUALITY OVERALL SUMMARY

The Quality Overall Summary (QOS) is a comprehensive summary or account of information, knowledge, and understanding of the drug substance/product, from its early development to commercialization, emphasizing what is critical for a robust, reproducible process as well as a consistent, reliable product quality (32). It is a formal template used to present relevant CMC information in a concise and organized manner to the FDA. It is the goal of the FDA reviewers that they receive summaries and conclusions based on the data obtained during the drug development process and not a "data dump" of all the raw data collected. The QOS is a document used to present the overall approach to acquiring product/process knowledge during the product development process. Additionally, the QOS document should describe the thought process explaining the science and risk-based rationales used in the decision-making process. It is a means to demonstrate concisely knowledge, understanding, and factors critical to the quality of the product and related supportive control strategies.

The QOS is also an integral part of the CTD, which is the basis for registration of human pharmaceuticals in Japan, Europe, and the United States (33). The ICH process has helped to harmonize the technical requirements for new drug submissions, and the CTD aims to harmonize the organization of the submissions. While each region has differing requirements for the first module of the CTD (Administrative Information and Prescribing Information), the rest of the modules contain the same information regardless of region. The QOS makes up the second module of the CTD (33). The QOS brings key issues and critical parameters of the product and process to the attention of the reviewer at the start of the review process. It differs from an expert report by emphasizing attributes and parameters that are critical to quality rather than just being a critical assessment of the product/process itself (34).

Although only summarized data and graphs are required for the QOS, all of the development experiments and studies must still be performed and the raw data stored at the plant product development center. It is likely that the FDA would inspect the raw data that support the QOS at the time of a PAI. It is important to note that this QOS does not reduce the number of research studies needed during drug development, it just reduces the amount and type of information the reviewer must study during the review of the application.

Use of the QOS offers many benefits to the pharmaceutical industry. It is a guide for the applicant in gathering, organizing, and presenting systematically expected CMC information essential to regulatory action. It focuses the presentation of scientific rationale emphasizing specific points leading to higher quality submissions and reviews. The QOS will assist in the early identification of potential issues or disagreement for quick resolution to achieve first cycle regulatory approval; in addition, it can facilitate a more relevant, focused scientific dialogue between the reviewer and the applicant. It is expected that the QOS will ultimately foster mutual confidence and trust between industry and regulators by sharing knowledge and understanding of the drug product and manufacturing processes, and eliminate the need to summarize standard CMC information by reviewers (32). The QOS also offers these other potential benefits:

- Steer development of potential "CMC regulatory agreement,"
- Serve as the primary review document,
- Serve as a concise, contemporary, and historical source of critical CMC information for FDA and the applicant,
- Guide and facilitate PAIs, and
- Expedite and streamline the CMC review process, thus enhancing review efficiency.

In order to obtain a truly comprehensive QOS, the emphasis must continue to shift from critical assessment to the assessment of those attributes and parameters that are critical to product quality. The use of summary tables, charts, and graphs is recommended rather than the inclusion of all the raw data. Thus, the QOS is anticipated to impact the conduct of PAI in several key ways. First, it is intended that the FDA will audit the raw data that exist in support of the QOS. Secondly, the QOS should help guide and facilitate the PAI because of the systematic approach it provides to the organization of CMC data.

DEVELOPMENT QUALITY SYSTEM

A development quality system defines the important quality requirements that enable the firm to adequately and sufficiently explore the key elements of design space (process parameters and material attributes) that impact the final quality of the product (35). The central goal of a quality system is ultimately to ensure consistent production of safe and effective products (35). A robust quality system promotes consistency by integrating effective knowledge-building mechanisms into everyday operational decisions. In this regard, the FDA is not focusing on the individual processes of drug manufacturing but rather on the attributes of the drug product itself.

The goal of a modern quality system is to promote continuous learning throughout the life cycle of drug development, both as assessed during the PAI and beyond, and to remove hurdles to continuous improvement following application approval. There are six basic elements in the quality system scheme for product development: quality assurance, facilities and equipment, material, production, packaging and labeling, and laboratory control.

Attributes of a robust quality system necessary for continuous learning throughout product life cycle are as follows (36):

- Science-based approaches,
- Decisions based on understanding product's intended use,
- Proper identification and control of areas of potential process weakness (including raw materials),
- Responsive deviation and investigation systems that lead to timely remediation,
- Sound methods for assessing risk,
- Well-defined and designed processes and products, from development through entire product life cycle,
- Systems for careful analyses of product quality, and
- Supportive management, both philosophically and financially.

Management of product development functions must play a key role in maintaining a robust quality system for PAI as well as future development because, as management, they are ultimately responsible for providing the leadership necessary to ensure that the quality system is in place, comprehensive, and well supported (37). Management must also ensure that the organizational structure is supportive of the quality system, ensuring that policies are in place and adequately resourced.

The quality system used in development has many of the same components as the production quality system with one major difference—how it handles change (38). While in production, the fixed manufacturing protocol is not intended to be varied in routine operation; in development, initial protocols are developed and require ongoing change in order to develop robust processes or methods. Although change is an inherent part of the development process, it cannot occur haphazardly, and must be handled thoughtfully. Certain manufacturing changes (e.g., changes that alter specifications, a critical product attribute or bioavailability, or design space) require regulatory filings and prior regulatory approval (35).

WITHDRAWAL OF CMC AND STABILITY GUIDANCE

In recent steps to further the implementation of cGMPs for the 21st Century, the FDA has withdrawn seven guidance documents. The reason for their withdrawal was that "some of the principles in these guidances are inconsistent with the agency's (cGMP) initiative" (39). The guidances being withdrawn are:

- Format and Content of the CMC Section of an Application (CDER, February 1987),
- Submitting Documentation for the Stability of Human Drugs and Biologics (CDER/CBER, February 1987),
- Stability Testing of Drug Substances and Drug Products (Draft) (CDER/ CBER, June 1998),
- Drug Product: CMC Information (Draft) (CDER/CBER, January 2003),
- Submission of CMC Information for Synthetic Peptides (CDER/CBER, November 1994),
- BACPAC I: Intermediates in Drug Substance Synthesis; Bulk Actives Post-approval Changes: CMC Documentation (CDER/CBER, February 2001), and
- Drug Substance: CMC Information (Draft) (CDER/CBER, January 2004).

According to the FDA's notice, "the withdrawals are part of a continuing review of all FDA guidances. We will continue to review our guidances for their consistency with the cGMP Initiative and may withdraw or revise other guidances if they do not reflect our current thinking or to align them with the concepts of the cGMP Initiative, the Quality by Design Initiative, or Question-based Reviews" (39). While the agency is developing new guidance documents to conform to these initiatives, the human drug pharmaceutical industry should consult ICH guidances (40).

CONCLUSION

The FDA pre-NDA approval inspection occurs at the intersection of product development activities and commercial manufacture, thus the inspection has the potential to assess both the product development process as well as the ability of the manufacturing component to begin commercial production. This fact has not changed since the PAI program's inception in the late 1980s and early 1990s. However, there have been many other changes since the program's inception some 15 years ago. These changes have been driven by the efforts of both the pharmaceutical and biotechnology industries and the FDA. One significant change for the better that has occurred over this time horizon is the improved preparation of the pharmaceutical and biotechnology industry for PAIs. This is the result of a significant amount of effort on the part of the industry to be ready for this type of inspection. This work has been exemplified by both short- and long-term efforts to be prepared for these critical inspections.

In the initial years of the program, many companies engaged in frantic, last-minute efforts to prepare for an FDA PAI. Thankfully, today that short-term frenzy has been replaced with longer-term preparation efforts that begin, in many cases, at the very early stages of the drug development process. Additionally, most firms have greatly enhanced their quality systems, which have also served to minimize the need for last-minute preparation efforts as well as substantially augment the internal quality assurance and quality control organizations. These improvements in the preparation efforts of the industry have not only served to enhance the likelihood that a firm will pass an FDA PAI; they have also served to facilitate the transition of a new product within a firm from product development to manufacturing.

Changes currently being proposed and formulated by the FDA have the potential to impact significantly PAIs of the future as well as the product development process itself. Two significant FDA changes are: (1) the recent modification of regulatory guidance derived from the GMPs of the 21st Century, and (2) the critical path initiative.

Recent regulatory guidance, such as cGMPs for the 21st Century, Quality Systems, QbD, and INDs—Approaches to Complying with cGMPs in Phase I, to mention a few, are indicative of the changing regulatory environment. As regulatory expectations continue to evolve and change over time through these and other initiatives, they will have a profound impact on the expectations of the agency at the time of a PAI. These changes coupled with the critical path initiative could have a significant impact on the development and regulatory oversight of new drug development. If the agency's efforts to lead the modernization of the drug development process are successful in concert with some of the initiatives reviewed here, we should see an enhanced ability of the industry to deliver safe and effective products to patients in a timely manner. This success will be good for the industry, regulators (FDA), and, most importantly, patients in need.

REFERENCES

1. Avallone H. Mid-Atlantic Region Pharmaceutical Inspection Program. 1990.
2. FDA. FDA Compliance Manual 7346.832 Pre-Approval Inspections. 1990.
3. Hynes MD, Medina C. Preapproval inspections: The critical compliance path to success. In: Medina C, ed. Compliance Handbook for Pharmaceuticals, Medical Devices, and Biologics. New York, NY: Marcel Dekker, 1998:463–491.
4. FDA. Bioresearch Monitoring: Good Laboratory Practice. Compliance Program 7348.808 Chapter 48. Federal Reserve, 2001.
5. CDRH/FDA. Good Laboratory Practice for Nonclinical Laboratory Studies (21 CFR 58). Federal Reserve, 2005.
6. FDA. Current Good Manufacturing Practice Regulations and Investigational New Drugs (63 FR 5233). 1998.

7. FDA. Guidance for Industry: E6 Good Clinical Practice: Consolidated Guidance, C. CDER. Federal Register, 1996. Available at: http://www.fda.gov/cder/guidance/959fnl.pdf.
8. FDA. Pharmaceutical cGMPs for the 21st Century: A Risk-Based Approach: Second Progress Report and Implementation Plan. 2003. Available at: http://www.fda.gov/cder/gmp/2ndProgressRept_Plan.htm.
9. FDA. IND Guideline for the Preparation of Investigational New Drug Products. 1991. Available at: http://www.fda.gov/cder/guidance/old042fn.pdf.
10. CDER/FDA. New Drug Evaluation: Pre-Approval Inspections/Investigations. Compliance Program 7346.832 Guidance Manual Chapter 46. Federal Reserve, 2005. Available at: http://www.fda.gov/cder/dmpq/7346-832-CDER.pdf.
11. FDA. Pharmaceutical cGMPs for the 21st Century: A Risk-Based Approach—Progress Report of the Working Group on Product Specialists on Inspection Teams. 2003. Available at: http://www.fda.gov/cder/gmp/2ndProgressRept_Plan.htm.
12. FDA. Pharmaceutical cGMPs for the 21st Century: A Risk-Based Approach: Final Report Fall 2004. Federal Reserve, 2004. Available at: http://www.fda.gov/cder/gmp/gmp2004/GMP_finalreport2004.htm.
13. FDA. Pharmaceutical Inspectorate (PI) Frequently Asked Questions, CDER. 2004. Available at: http://www.fda.gov/cder/gmp/PI-q&a.htm.
14. FDA. Innovation and Continuous Improvement in Pharmaceutical Manufacturing—Pharmaceutical CGMPs for the 21st Century. Federal Register, 2004. Available at: http://www.fda.gov/cder/gmp/gmp2004/manufSciWP.pdf.
15. FDA. Risk-Based Method for Prioritizing CGMP Inspections of Pharmaceutical Manufacturing Sites—A Pilot Risk Ranking Model. 2004. Available at: http://www.fda.gov/cder/gmp/gmp2004/risk_based.pdf.
16. Nasr MM. A New Pharmaceutical Quality Assessment System (PQAS) for the 21st Century: Why is it needed, what does it mean, and how do we get there? In: AAPS Workshop. North Bethesda, MD; 2005.
17. Migliaccio G. ICH Quality Systems (AKA Q10). In: AAPS Workshop—Pharmaceutical Quality Assessment. North Bethesda, MD; 2005.
18. CDER/FDA. ONDC's New Risk-Based Pharmaceutical Quality Assessment System. 2004. Available at: http://www.fda.gov/ohrms/dockets/ac/06/briefing/2006-4241B1-02-11-FDA-QbD%20ONDC%20White%20Paper.pdf.
19. FDA. ICH Q8 Pharmaceutical Development. Federal Register, 2005.
20. CDER/FDA. Guidance for Industry: PAT: A Framework for Innovative Pharmaceutical Development, Manufacturing, and Quality Assurance. 2004. Available at: http://www.fda.gov/cder/guidance/6419fnl.htm.
21. FDA. Innovation or Stagnation: Challenges and Opportunity in the Critical Path to New Medical Products. 2004. Available at: http://www.fda.gov/oc/initiatives/criticalpath/whitepaper.html.
22. Woodcock J. Pharmaceutical quality in the 21st century: An integrated systems approach. In: AAPS Workshop on Pharmaceutical Quality Assessment: A Science and Risk-based CMC Approach in the 21st Century. North Bethesda, MD; 2005.
23. FDA. FDA Issues Advice to Make Earliest Stages Of Clinical Drug Development More Efficient. 2006. Available at: http://www.fda.gov/bbs/topics/news/2006/NEW01296.html.
24. FDA. TITLE 21—Current Good Manufacturing Practice for Finished Pharmaceuticals: Food and Drugs; CHAPTER I—FDA Department of Health and Human Services; Subchapter C—Drugs: General (21 CFR 211). Federal Register, 2006.

25. Shuren J. Current Good Manufacturing Practice Regulation and Investigational New Drugs. Federal Register, 2006; 71(10):2458–2462.

26. FDA. Guidance for Industry: INDs—Approaches to Complying with CGMP During Phase 1. Federal Register, 2006. Available at: http://www.fda.gov/cber/gdlns/indcgmp.htm.

27. FDA. FDA Guideline on the Preparation of Investigational New Drug Products (Human and Animal). 1991. Available at: http://www.fda.gov/cder/guidance/old042fn.pdf.

28. FDA. Current Good Manufacturing Practice Regulation and Investigational New Drugs: Companion Document to Direct Final Rule. Federal Register, 2006: 2494–2496.

29. Simmons JE. Understanding key terms for modern quality assessment. In: AAPS Workshop on Pharmaceutical Quality Assessment: A Science and Risk-Based CMC Approach in the 21st Century. North Bethesda, MD; 2005.

30. CDRH/FDA. Current Good Manufacturing Practice for Finished Pharmaceuticals (21CFR211.165). Federal Regulations, 2005.

31. Shuren J. Submission of Chemistry, Manufacturing, and Controls Information in a New Drug Application Under the New Pharmaceutical Quality Assessment System; Notice of Pilot Program. Federal Register, 2005; 70(134):40719–40720.

32. Poochikian G. QOS: FDA Perspective. In: AAPS Workshop on Pharmaceutical Quality Assessment: A Science and Risk-based CMC Approach in the 21st Century. North Bethesda, MD; 2005.

33. FDA. Guidance for Industry M4Q: The CTD–Quality. 2001. Available at: http://www.fda.gov/cder/guidance/4539Q.htm.

34. McArdle JV. The comprehensive QOS: Industry perspective. In: AAPS Workshop on Pharmaceutical Quality Assessment: A Science and Risk-Based CMC Approach in the 21st Century. North Bethesda, MD; 2005.

35. FDA. Guidance for Industry: Quality Systems Approach to Pharmaceutical Current Good Manufacturing Practice Regulations—DRAFT. Federal Register, 2004. Available at: http://www.fda.gov/cder/guidance/7260fnl.htm.

36. Friedman RL. Continuous Improvement under Modern Quality Systems and CGMPs. 2005. Available at: http://www.fda.gov/cder/present/DIA2005/Friedman.pdf.

37. FDA. Guidance for Industry: Quality Systems Approach to Pharmaceutical Current Good Manufacturing Practice Regulations. 2004. Available at: http://www.fda.gov/ohrms/dockets/ac/05/briefing/2005-4136b1_05_pharmaceutical%20CGMP.pdf.

38. Hynes MD. Developing a strategic approach to preparing for a successful pre-NDA approval inspection. In: Hynes MD, ed. Preparing for FDA Pre-approval Inspections. New York, NY: Marcel Dekker; 1999:10–30.

39. Administration, U.F.A.D. Guidance for Industry on Chemistry, Manufacturing, and Controls Information: Withdrawal and Revision of Seven Guidances. Federal Register, 2006; 71(105):31194–31195.

40. McCormick D. FDA withdraws seven CMC and stability guidance documents. Pharm Technol. 2006; 30(7):18–20.

2

FDA's Risk-Based Approach to Inspections

Ronald F. Tetzlaff

Parexel Consulting, Alpharetta, Georgia, U.S.A.

Jeanette M. Buckwalter

Eli Lilly and Company, Indianapolis, Indiana, U.S.A.

INTRODUCTION

In August 2002, the Food and Drug Administration (FDA or agency) issued a press release announcing its initiative, "Pharmaceutical cGMPs for the 21st Century: A Risk-Based Approach" (1). When the agency rolled out this initiative, Dr. Lester Crawford (the FDA's then acting commissioner) predicted that "in two years you won't recognize the FDA." Since then, the FDA has made a number of fundamental changes in the way it regulates the pharmaceutical industry.

To understand why the FDA launched the initiative, it may be instructional to consider some of the factors that influence the pharmaceutical industry's ability to get new drugs to the market and to sustain competitive profitability levels. Prior to the FDA launching its initiative in 2002, the agency's priorities for pharmaceutical manufacturing focused heavily on enforcing good manufacturing practices (GMPs), performing pre-approval inspections (PAIs), and reviewing post-approval changes for new drugs and biologics. The GMP Regulations (21 CFR parts 210 and 211) have remained substantially unchanged for more than 25 years despite technological innovations of great proportions and significant updates in global regulatory requirements.

The GMPs for the 21st Century initiative was a proactive program intended to foster innovation in the pharmaceutical industry (2). The reasons for the FDA adopting a risk-based approach to pharmaceutical manufacturing are rooted in a decline in new drug approvals to record levels. In addition, the number of drug GMP inspections by the FDA had been steadily declining for more than two decades, while the number of registered drug and biologic manufacturers had increased approximately sevenfold during the same period. The FDA was concerned that innovations in the pharmaceutical industry were not keeping pace with technological developments and scientific breakthroughs.

Recent Drug Development Trends Lead to Changes in FDA Regulation of Drug Manufacturing

A brief review of drug development trends in recent years provides an insight into why the FDA is fundamentally changing the way it will regulate pharmaceutical manufacturing during the next decade. During the past decade, the FDA and other regulatory authorities have seen a dramatic decline in the number of applications for new molecular entities (NMEs) and biologic license applications (BLAs). For example, the number of BLAs received by the FDA decreased by about sevenfold from FY 1993 to FY 2003, and the FDA received about 50% fewer applications for NMEs in FY 2003 than during the mid-1990s (3). Of the drugs approved by the Center for Drug Evaluation and Research (CDER), 36 had novel structures, classified as NMEs or new BLAs, up from 21 in 2003. More and more mergers are leading to fewer companies that are capable of developing NMEs, and fewer blockbuster NMEs are in the pipeline. A number of blockbuster drugs have or will shortly come off patent, which increases pressures for affected companies to find new revenue sources or to improve operational efficiencies to compensate for the market share that is being lost to generics. The decline in new drug approvals is contrasted by generic drugs, which have experienced a steady increase in the number of abbreviated new drug applications. For example, the number of approvals increased from 307 in FY 2001 to 563 in FY 2004.

While the pharmaceutical industry continues to increase spending for research and development (R&D), the total number of new drugs approved by the FDA has declined markedly. Prior to the GMPs for the 21st Century initiative, the FDA's actions and programs did not encourage companies to invest in changes that would result in reliable innovation and investments. The FDA's approval rate for new drugs was at its lowest level in a decade. For example, the CDER approved 17 NMEs in FY 2002 versus 53 approvals in FY 1996 (4).

The overall success rate for new drug development decreased from about 14% during the 1980s to about 8% during FY 2004. In addition, the industry experienced a decline in the number of submissions of novel product applications during the past decade. For example, during the mid-1990s, the CDER handled about 44 applications for NMEs per year, and the Center for Biologics Evaluation and Research (CBER) about 44 BLAs per year. In FY 2004, CDER received 28 applications for NMEs and the CBER received 20 BLAs.

Since the passage of the Prescription Drug User Fee Act (PDUFA) (6) in 1992, the FDA has reduced its overall time for application approvals. For example, in FY 2004, FDA approvals met the PDUFA targets. During FY 2003 and FY 2004, the median approval time for priority NMEs was 6 months, and for BLAs, the median approval time was 15.6 and 14 months, respectively. In 2004, the CDER received 108 new drug application (NDA) submissions versus 110 in 2003. The CDER approved 29 drugs under the priority review program and 90 drugs under its standard review program. The CDER data for 2004 showed that median approval times dropped for priority drugs (i.e., those having potential for significant advances over existing treatments), and standard NDAs, and approval times for the BLAs have continued to drop.

Nearly 50% of new drugs that reach the final stages of Phase 3 studies are not approved because of inadequate evidence of safety and effectiveness. During the past decade, a number of scientific breakthroughs in the biotech industry (genomics, proteomics, medical imaging, and nanotechnology) were expected to help fill new drug pipelines, but these expectations have not been realized.

At a time when new drug approvals are declining, R&D spending by member companies of Pharmaceutical Research and Manufacturers of America (PhRMA) has reached record levels. PhRMA reported a 12% increase in R&D spending to $38.8 billion in 2004 compared with 2003 (7). Figure 1 shows that during the 10-year period from 1993 to 2003, the R&D spending for biomedical research more than doubled. During the same time period, the number of major drug and biological submissions decreased by approximately 50% as depicted in Figure 2 (3).

The number of NME applications submitted to the FDA by the pharmaceutical industry has declined in recent years. The 10-year trend is shown in Figure 3.

The R&D cost of developing an NME has risen to all-time highs (as much as $1 billion or more), which severely limits the number of companies with the financial wherewithal to test new compounds for safety and effectiveness. In the past, the FDA did not concern itself with the costs associated with new drug development, leaving drug companies on their own to deal with the financial considerations. The FDA's "Critical Path" report acknowledges some of the daunting challenges associated with establishing safety and effectiveness of promising new drugs while keeping costs as inexpensive as possible. Never before has the agency been so willing to collaborate with drug developers, academia, and patient advocate groups to find ways to get promising new therapies to patients faster while increasing predictability and reducing costs.

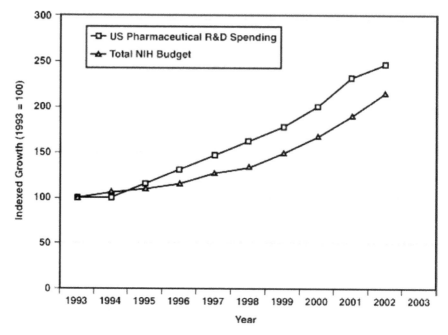

Figure 1 Ten-year trends in biomedical research spending, as reflected by the NIH budget (Budget of the United States Government, appendix, FY 1993–2003) and by pharmaceutical companies' R&D investment (PAREXEL's Pharmaceutical R&D Statistical Sourcebook 2002/2003). *Abbreviation*: R&D, research and development. *Source*: From Ref. 3.

According to a *Wall Street Journal* article, the top 16 pharmaceutical companies spent $90 billion in 2001 on manufacturing operations (labor, materials, operations, and depreciation). Manufacturing costs represent about 36% of their total expenditures, and this is more than twice their cost of R&D (9). Manufacturing costs are receiving more attention by pharmaceutical and biotech manufactures as profit margins have begun to decline in recent years. The pharmaceutical and biotech sectors generally have not experienced the double-digit growth and profitability levels that were so characteristic of this industry a few years ago. Manufacturing issues were a top reason for delays in approvals of NMEs reviewed by the CDER. For CBER products, the top reason for delayed approvals of BLAs in FY 2001 was changes to the manufacturing process.

In recent years, the pharmaceutical industry experienced a shift in the number of new drugs in the pipeline, and the mix between pharmaceuticals and biotech products has changed. A decade ago, pharmaceutical manufactures were averaging about 60 new drug and four biotech approvals per year. By year 2002,

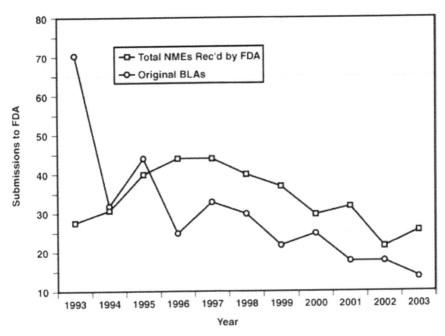

Figure 2   Ten-year trends in major drug and biological product submissions to the FDA. The figure shows the number of submissions of NMEs—drugs with a novel chemical structure—and the number of BLA submissions to FDA over a 10-year period. Similar trends have been observed at regulatory agencies worldwide. *Abbreviations*: NME, new molecular entities; BLA, biologics license application. *Source*: From Ref. 3.

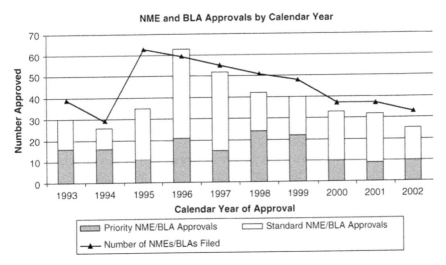

Figure 3   NME and BLA approvals by calendar year (1993–2002). *Abbreviations*: NME, new molecular entity; BLA, biologic license application. *Source*: From Ref. 8.

the pharmaceutical and biotech sectors each had approximately 20 approvals per year.

FDA Changes Approach to Inspections to Reflect Risk-Based Model

Another consequence of the GMPs for the 21st Century initiative has been the changes in the FDA's approach to inspections, including the application of risk-based principles to allow better management of agency obligations with fewer resources (e.g., allowing fewer inspections at targeted, higher risk facilities). The FDA has made basic changes to its inspection approach and enforcement posture (e.g., fewer Warning Letters and certain sanctions have declined), and it is promoting novel approaches for quality management. These changes have allowed unprecedented opportunities for companies to improve their quality management programs in ways that will increase assurances of product quality and safety while reducing costs and the time it takes to get their products to the market.

The majority of the changes that have taken place so far under the FDA's risk-based initiative have focused around better decision-making by the industry and the FDA. The FDA hopes to stimulate innovations in the industry and to foster a science-based approach during its review of NDAs and during GMP inspections of manufacturers. A key FDA objective is to encourage use of sound science to better understand sources of variability during product development and manufacturing. Through better management of the safety of marketed drug products, a number of benefits are expected (e.g., focusing on high risk issues should reduce the time needed to conduct FDA inspections, and should improve the consistency and predictability of the FDA's actions, expectations, and decisions). The FDA is focusing attention on management's accountability for quality (preventative and corrective actions). The FDA appears committed to allowing companies greater flexibility under the risk-based paradigm as evidenced by several significant policy changes and a number of recent guidance documents that reflect its new ways of thinking.

Drug development costs are heavily impacted by manufacturing issues associated with developing a scaleable process and ensuring conformance to cGMP during the production of clinical trial materials and scale-up to commercial batches. Since GMP regulations have not been substantially changed in more than two decades, the FDA is considering improved ways to ensure that new efficiencies gained by adoption of innovative predictive tools will be synchronized with GMP requirements. While changes have not yet been made to the drug GMPs (Parts 210 and 211), the agency seems willing to revise the regulations where justified by new innovation [i.e., if they will result in more safe and effective drugs making it to the market (or faster)]. The agency appears intent on providing pharmaceutical companies with greater freedom to manage post-approval changes if they apply scientifically sound, risk-based approaches to understanding their processes and controlling the parameters that cause variability.

For companies that demonstrate the ability to successfully manage risks, the FDA is likely to give up much of the review functions they have insisted upon in the past for NDAs. If the FDA has assurance that companies are managing risk, it will be supportive of shifting a significant portion of their application reviews to pre-approval and post-approval inspection audits to confirm that changes have been thoroughly evaluated to support the change. The FDA now recognizes that modern quality systems can produce increased understanding of the process and product, and by using this knowledge, manufacturers should be given greater freedom to make improvements without always having to obtain prior approvals from the FDA. Ultimately, this increased knowledge gained from effective quality systems should result in fewer (less frequent) and shorter inspections by the FDA (once they gain confidence).

The FDA recognized the need to modernize the rules under which pharmaceuticals are manufactured, and wanted to overhaul its GMP programs to improve product quality, encourage development of innovative technologies, and reduce the development and manufacturing costs for drug products. On August 21, 2002, the FDA issued a press release to announce a new two-year initiative, "Pharmaceutical cGMPs for the 21st Century: A Risk-Based Approach" (2), and the details were outlined in a concept paper, "A science and risk-based approach to product quality regulation incorporating an integrated quality systems approach" (2). The agency simultaneously issued a letter to all FDA employees together with Questions and Answers about the initiative (10).

While the agency had announced projects and initiatives in years past, none has had a greater impact on the way the FDA manages the drug inspection and approval processes. In the past, most major initiatives were triggered by an adverse event that prompted the FDA to take action in reaction to that event (such as the generic drug scandal that resulted in the PAI program). The "GMPs for the 21st Century" initiative is unique in that there was not a single adverse event that triggered this initiative, but rather the agency reacted proactively to the declining number of new drug approvals. Another key difference from previous initiatives was the fact that this program was rolled out, not as a pilot project, but rather represented a fundamental new approach to the way the agency intends to manage GMPs following a risk-based model.

When the FDA launched its GMPs for the 21st Century initiative, it intended to fundamentally change the new drug approval process and its regulation of pharmaceutical manufacturing. The agency was clearly committed to facilitating drug development and wanted to introduce new ways of doing business that would make it more predictable and encourage innovation by industry. Among the planned changes were increased collaborations with drug sponsors, revised regulations, and issuance of new guidance documents to address current new drug development issues. Substantial headway has been made toward developing a new paradigm for applying risk management to drug safety and quality management to drug manufacturing. The FDA has revised or withdrawn many guidance documents that had previously been published, and has issued a number of new guidance documents that are designed to encourage companies to use new state of the art manufacturing equipment and novel methods to assure product quality. The following is a brief description of some of the key changes that have taken place since the August 2002 announcement of the GMPs for the 21st Century initiative.

The agency has changed its posture, and is now encouraging the industry to develop more sophisticated tools for predicting safety, effectiveness, and commercial scale-up, and is looking for ways to help companies find more efficient and cost-effective techniques to develop promising compounds. For example, the FDA encourages predictive screening techniques to allow rejection or redesign of products before incurring the costs of clinical trials including: 1) application of genomic and proteomics techniques to assess safety of new biomaterials; 2) improved computer and animal models to predict safety and effectiveness (such as biomarkers, and surrogate end-points); and 3) use of more modern engineering studies and scientific support for manufacturing processes.

The FDA is conducting a number of studies to assess the decline in new drugs under development, including a retrospective root cause analysis of the reasons for nonapproval by during its first review cycle, and two three-year pilot studies that will identify ways that FDA reviewers might collaborate with drug sponsors to shorten approval times for products with fast-track designation (e.g., computer modeling, biomarkers, validated assays, and surrogate end-points for clinical trials).

While the FDA is publicly advocating use of new predictive techniques and tools, most companies are reluctant to make investments when outcomes are uncertain. Until predictive tools have gained overall acceptance by the FDA and other regulatory agencies, many companies are taking a "wait and see" approach. Even though the return-on-investment would be large if companies could reduce the size of clinical trials or reject candidate compounds before the clinical trial phases, the added financial burden of proving a new approach is a risk many companies are reluctant to take at this time. To assist the industry in overcoming the obstacles to developing new predictive tools, the FDA intends to collaborate with the National Institutes of Health, academicians, and companies in the private sector.

In summary, drug development costs are heavily impacted by manufacturing issues associated with developing a scaleable process and ensuring conformance to cGMPs during production of clinical trial materials. While the FDA's GMP regulations have not been substantially changed in more than two decades, on December 4, 2007, the FDA withdrew a 1996 proposed rule and published a direct final rule as a first increment of future modifications to those regulations (11). These changes were made to not only modernize and clarify the requirements but also to harmonize some of the regulations with respect to other international agencies. Several sections of 21 CFR Parts 210 and 211 were changed including aseptic processing, equipment cleaning, and product containers and closures (12). The FDA is looking into ways to ensure that new efficiencies gained by the adoption of innovative predictive tools will be synchronized with GMP requirements. Finally, it appears that under the GMPs for the 21st Century initiative, the agency intends to give pharmaceutical companies greater freedom to manage post-approval changes if they apply scientifically sound, risk-based approaches to understanding their processes and controlling the parameters that cause variability. If companies can successfully mange risks, then the FDA seems intent on giving up much of the pre-approval application review function it has insisted upon in the past, and will be moving toward post-approval inspection audits to confirm that changes have been thoroughly evaluated. The FDA now recognizes that modern quality systems can be used to increase an understanding of the product and its processes, and using newly gained knowledge, manufacturers should be given the freedom to make improvements without having to obtain prior approval from the FDA. Ultimately, this increased knowledge derived from effective quality systems should result in fewer (less frequent) and shorter inspections by the FDA (once it gains confidence).

A number of these changes are outlined in this chapter.

FDA ADOPTS A RISK-BASED APPROACH TO PHARMACEUTICAL QUALITY

FDA's GMPs for the 21st Century Initiative

On August 22, 2002, the FDA launched its initiative, "Pharmaceutical cGMPs for the 21st Century: A Risk-Based Approach" (1). The FDA's goal is to modernize its regulation of pharmaceuticals through: 1) an overhaul of its GMP programs, 2) development of innovative technologies, and 3) a reduction in the cost of development and manufacturing. The agency is making a paradigm shift in its new drug approval process and is reevaluating its enforcement of GMP regulations. The FDA appears committed to finding ways to promote drug development, and is willing to allow companies to use improved quality management approaches that foster innovation by industry. The agency wants to be seen as being more predictable. Examples of changes that have emerged include, but are not limited to:

- The FDA has expanded its collaborations with drug sponsors.
- The FDA is reviewing current regulations and has issued a number of new guidance documents to address current new drug development issues.
- The FDA is encouraging companies to apply risk management to drug safety and quality management to drug manufacturing.
- The FDA is encouraging industry to develop more sophisticated tools for predicting safety and effectiveness, and to facilitate commercialization of promising compounds.

- The FDA is conducting a number of studies to assess the decline in new drugs under development.
- The FDA is performing retrospective root cause analysis of the reasons for nonapproval during a first review cycle.
- Pilot studies are being designed to identify ways that FDA reviewers might collaborate with drug sponsors to shorten approval times for products with fast-track designation (e.g., computer modeling, biomarkers, validated assays, and surrogate endpoints for clinical trials).

Beginning in FY 2005, the FDA began using the risk-based approach across all five centers to decide the inspection work plan (i.e., prioritization of inspections), including assessments to identify the internal and external hazards that companies face, and to estimate the risks and characterization of the hazard. In deciding whether to inspect each company or the frequency of inspections, the FDA is considering the public health consequences in a more systematic manner using science-based approaches. Criteria now being applied include: the nature and type of prior violations, annual production volume, profile classes of products produced, complexity of manufacturing processes and controls used to minimize risks, probability and severity of harm, vulnerability of product to contamination by pathogens or deliberate tampering, and overall health and safety consequences. The FDA is using a semi-quantitative method for establishing a site risk potential (SRP) value. The SRP is a weighted calculation based on the individual risk potentials for the process, product, and facility (each scored as high, medium, or low). The facility weighting factor takes into account whether the facility has been recently inspected, the number and severity of previous GMP deviations, and the volume of products produced in the facility. The product weighting factor is based on dosage form and route of administration (e.g., sterile versus topical), prescription versus over-the-counter (OTC) status, or products that have had a higher frequency of serious recalls. The process weighting factor takes into account the complexity of the manufacturing process and/or whether or not the product is prone to contamination during processing.

Under the new risk-based work planning, some companies will see fewer FDA inspections than in the past, but others may experience more frequent visits and/or visits of longer duration. For example, a company that manufactures only a single OTC product and has no significant history of prior GMP deficiencies is likely to be inspected less frequently than in the past, whereas, a large multiproduct producer of sterile injectable drugs with a history of compliance issues should expect increased scrutiny (i.e., inspections at more frequent intervals and in greater depth).

FDA's Critical Path Report

In March 2004, the FDA published its "Critical Path" report that was intended to help promote a transformation of the drug development process (13,14). The agency acknowledges its share of responsibility for the development of novel drugs and is soliciting the assistance of industry, academia, and other federal agencies. The agency is studying the reasons for the decline in the number of new drugs in development, including the gap between basic science and applied science. It recognizes the need to improve scientific insights when evaluating the safety and efficacy of a new compound. Many drug development failures could be avoided if the industry had better tools and scientific models for dealing with uncertainties. The Critical Path initiative encourages the development of new predictive tools to allow screening of compounds to identify probable failures earlier in the development cycle. The FDA is encouraging development of breakthrough technologies, and plans collaborative research to enable creation of a new generation of medical products.

Under the Critical Path initiative, the FDA intends to assemble appropriate data and information and to establish correlations and to formulate new generalized principles to enable faster drug development with more reliable results in predicting drug safety and efficacy. The Critical Path public docket closed on July 30, 2004, and on March 20, 2006, the FDA published its "National Critical Path Opportunities List," which is an effort to determine how the agency, industry, and academia can cooperate most effectively (15). This list identifies 76 priority issues for product development for which innovative approaches and emerging technologies might produce the most benefits. The ultimate intent is to gain collaboration between the FDA, other government agencies, academia, and pharmaceutical companies to develop an innovative path toward more efficient drug development, leading to more new drug approvals in a shorter time. The goal is to shorten the gap between the rapidly emerging biomedical discoveries and the development of safe and effective therapies. The ultimate goal is to modernize the drug development process by 2010 to allow medical discoveries to help patients sooner and at a lower cost.

Post-Marketing Drug Safety/Pharmacovigilance

The FDA is encouraging companies that market new drugs to apply sound scientific principles during review and evaluations of post-marketing experiences. The agency is giving greater attention to addressing managements' reviews of safety hazards throughout the lifecycle of a product. It expects companies to establish scientifically sound methods and tools for its pre-marketing assessments and to follow plans for minimizing risk through effective monitoring systems. In March 2004 and May 2005, the FDA published four draft guidance documents that describe new concepts and principles that are going to fundamentally change the approaches used to demonstrate drug safety. In the past, companies and the FDA largely managed risk for marketed products by the statements made in label inserts concerning indications for use, and contraindications and warnings. The FDA is now encouraging companies to perform controlled, long-term safety studies, and to employ data mining techniques to establish linkage between adverse experience and a drug and/or improved systems for monitoring performance of drugs over time.

A number of widely prescribed drugs have been voluntarily removed from the market in recent years because of post-marketing data that raised questions about the safety of the marketed products. In a number of these instances, post-marketing information obtained from complaints, adverse experiences, or Phase IV studies revealed new side effects or other adverse safety data (subsequent to FDA approval of the NDAs).

On March 24, 2004, the FDA published three new guidance documents that describe some new principles and approaches that may be of use in improving the safety of certain high-risk drugs and biological products (16–18). In May 2005, it issued the draft guidance "'Drug Watch' for Emerging Drug Safety Information" (19). These four new guidance documents address management of safety hazards throughout the lifecycle of a product, including methods for pre-marketing assessments, plans for minimizing risk through monitoring systems, and concepts and practices for pharmacovigilance/pharmacoepidemiology.

While these draft guidance documents are limited in scope to certain products, and are subject to further revisions as the industry and academia provide feedback about their experiences, the basic principles advocated in these guidance documents will fundamentally change the approaches that can be successfully used to demonstrate drug safety. The risk assessment approach advocated by the FDA goes beyond the current system of managing risk by statements in label inserts. For example, the agency is advocating use of controlled, long-term safety studies that include use of data mining to establish linkage between adverse experience and a drug. The FDA expects companies to improve their systems for monitoring the performance of drugs over time, and is encouraging sponsors to use risk assessment approaches during Phase 3 studies.

The four draft guidance documents outline a number of key issues to be addressed during the pre-approval safety reviews. One example of a new risk-based approach to drug safety is a tool, RiskMAP, which is intended to prevent serious adverse events and to increase the probability that a drug will provide valuable benefits. Finally, the FDA realizes that it is not possible to identify all safety issues during clinical trials. Post-marketing safety data collection and assessments are receiving greater attention by the agency, and its draft guidance identifies a number of key factors to be addressed to allow more informed decisions during pharmacovigilance and pharmacoepidemiological studies. Finally, the FDA plans to develop improved systems and processes for disseminating to healthcare professionals and patients emerging drug safety information about marketed drug products via the Internet. On a new FDA Web page called the "Drug Watch," the agency will identify drugs that it is currently evaluating for early safety issues (i.e., signals) to assess the significance of emerging data and information about the safety of marketed products. This draft guidance describes information that will be posted on the Web site, criteria for inclusion on the list (and removal), procedures for notifying sponsors, and impact on drug promotions and advertising (20).

FDA's Quality Systems Approach to Pharmaceutical GMPs

The current GMP Regulations (21 CFR parts 210 and 211) have not been changed substantially since they were promulgated in September 1978 (21). On May 3, 1996, the FDA published a proposed rule in the Federal Register that included a number of significant revisions to the drug GMPs (22). The agency has had comments from the industry under advisement, but has neither finalized nor withdrawn the proposed rule. It appears that the FDA does not intend to finalize the May 1996 proposed rule, but favors a general overhaul of the regulations (currently in progress as an internal agency project). The FDA has not announced when it intends to revise the GMP regulations, but when it does; it is highly likely that any proposed rule will focus on risk-based principles.

The FDA published a guidance document in September 2006 entitled, "Quality Systems Approach to Pharmaceutical Current Good Manufacturing Practice Regulations" (23). The agency considers this document as being a bridge in the gap between the 1978 regulations and current thinking about quality management concepts. The FDA recognizes that the GMPs for finished pharmaceuticals do not address current expectations for quality management, quality by design, and other risk-based approaches to quality.

This guidance describes a quality systems model that will allow manufacturers to implement modern quality systems that conform to cGMPs. A key concept that is being forwarded by the FDA is that companies that apply sound scientific principles to understand their processes and sources of variability should be allowed to make improvements without the need for prior approval supplements.

While this document represents guidance and the language is not legally binding as with cGMP requirements, a number of the sections describe principles and practices that are cGMP. Indeed, a number of the sections cross-reference the applicable sections of Part 211. The FDA clearly intends to harmonize the cGMPs with other widely recognized systems for quality management, ISO 9000 (24), International Conference on Harmonization of Technical Requirements for Registration of Pharmaceuticals for Human Use (ICH) quality management principles (e.g., ICH Q8, ICH Q9, and ICH Q10) (25), and the FDA's Quality Systems Regulations (QSR) (21 CFR Part 820) (26).

Quality management concepts and principles are evolving and are converging across international borders and across product types. While it may be some time before the FDA codifies its quality management and risk-based approaches as requirements in revisions to the GMP regulations, prudent manufacturers will, where appropriate, reevaluate their existing quality systems in light of the current thinking of agency officials. The following are some examples of key areas where the agency is advocating application of risk management principles, including, but not limited, to the following:

- The FDA is shifting its position on the management of post-approval changes for approved applications. The agency now recognizes that when companies use risk-based approaches to understanding their processes/products and the sources of variability, the industry can manage post-approval changes (i.e., a reduced level of prior approval review by the agency). By giving the industry the option to manage post-approval changes, the FDA intends to shift its focus and priorities on audits to confirm the appropriateness/adequacy of data used to support changes.

- The FDA is encouraging companies to utilize risk-based approaches and innovative technologies [such as the process analytical technology (PAT)] to increase their understanding of their products and processes.
- The FDA is insisting that companies apply sound risk management principles throughout product lifecycles to achieve continuous improvement and adopt innovations where feasible.
- The FDA is asserting that quality systems be in harmony with quality management principles and be designed to take into account the specific conditions at each manufacturer including an integrated approach to such factors as organizational structure, responsibilities, procedures, processes, and resources.
- The FDA is advocating that risk management principles be applied to the setting of specifications and process parameters.
- The FDA is ensuring that quality management when applied to discrepancy investigations and corrective actions will result in sound decisions about the disposition of affected products.

The FDA's quality systems model focuses on four key elements as follows:

- Management responsibilities
  - Provide leadership
  - Structure the organization
  - Build quality systems to meet requirements
  - Establish policies, objectives, and plans
  - Review the system
- Resources
  - Provide general arrangements
  - Develop personnel
  - Provide facilities and equipment
  - Control outsourced operations
- Manufacturing operations
  - Design and develop product and processes
  - Monitor packaging and labeling processes
  - Examine inputs
  - Perform and monitor operations
  - Address nonconformities

- Evaluation activities
  - Analyze data for trends
  - Conduct internal audits
  - Perform risk assessment
  - Initiate corrective actions
  - Take preventative action
  - Promote improvements

FDA RISK-BASED INSPECTION APPROACH

While the quality systems inspection approach rolled out as part of the GMPs for the 21st Century initiative included some new compliance programs, systematic inspections of quality systems are certainly not new to the FDA. For several decades the agency applied a systematic approach to evaluating industries when problems were known or suspected (such as during the septicemia outbreaks in the large volume parenteral industry during the 1970s and the team biologics initiative that was launched in the late 1990s) (27). A formal quality system inspection program was first implemented by the Center for Devices and Radiological Health (CDRH) as part of the implementation of the QSR for medical devices, and the CBER has followed suit and issued its own version of a quality systems approach for biological drug products. The following describes the current inspection programs for PAIs and risk-based inspection programs in the CDER, the CBER, and the CDRH.

PAI Program

Under the FDA's risk-based initiative, the basic inspection approach for preapproval inspections (PAIs) is similar to that used in the past, but the agency will consider risk-based factors when deciding which sites to inspect, the inspection frequency, and the systems to be covered. FDA investigators may elect to follow any number of approaches for conducting inspections of pharmaceutical manufacturers. As in the past, the PAIs will focus on: 1) verifying integrity of information supporting applications (e.g., made to the FDA), and 2) to determine conformance with GMPs. FDA inspections include a review and evaluation of the facilities, processes, and the product(s) to determine whether there are adequate assurances of identity, strength, quality, and purity.

The FDA revised the PAI program in September 2003 and again in March 2004 (28). The list of narrow therapeutic range drugs was deleted, and items that are no longer required to be submitted in the application were identified. Instead, the agency elected to have these items evaluated by the FDA investigator during PAIs (except for sterile product validation and associated information that the FDA still expects companies to file in the application for CDER drug products) (29). In addition, the program not only reduced the number of inspection categories, but also reduced the number categories requiring mandatory inspection. The program gives greater discretion to Field Offices to decide whether a PAI is warranted.

The following categories will prompt a pre-approval or cGMP inspection:

- NMEs [both active pharmaceutical ingredients (APIs) and finished dosage forms]
- Priority NDAs
- First application filed by applicant
- For-cause inspections
- For current applications—if the current cGMP status is unacceptable or greater than 2 years
- For certain pre-approval supplements, such as site changes or major construction, if the cGMP status is unacceptable
- Treatment IND inspections
- The CDER has information that suggests clinical supply inspection is needed to protect public health.

The following categories may prompt a PAI or cGMP inspection on the basis of the discretion of the FDA district office:

- All original applications not listed above
- All pre-approval supplements not listed above.

In addition, district offices may use their discretion to decide whether to conduct a PAI in response to 10-day status report for changes being effective (CBE) supplements or other situations besides specific assignments from headquarters.

CDER Risk-Based Inspection Programs

Following a successful pilot program in 2001, the FDA revised its Compliance Program (CP 7356.002) for Drug Process Inspections in February 2002 to introduce a new inspection approach. Under this program, the FDA conducts inspections to assess a firm's state of control (i.e., whether it "produces finished drug products for which there is an adequate level of assurance of quality, strength, identity and purity") (30). The quality systems inspection represents more of a "top-down" rather than the traditional approach used in the past (which was a "bottom-up" approach). While the new quality systems approach is intended to focus on policies and procedures that define quality management programs and the ability to maintain effective controls, the FDA will continue to focus on deviations and exceptions that become apparent during the inspections.

This new program is intended to reduce inspection duration and/or eliminate the need for separate inspections of individual profile classes. One of its most significant features is the fact that if the FDA finds one quality system to be not operating under a state of control, the FDA may deem all profile classes to be in noncompliance (and subject to regulatory sanctions). Conversely, a single inspection that is found to be in compliance can be used as the basis for establishing satisfactory compliance for all profile classes (which eliminates the need for individual inspections based on profile classes). The changes under this program will allow the FDA to perform fewer inspections and reduce the total time spent performing inspections. Companies that have adopted effective quality systems should benefit from fewer inspections of shorter duration, but companies with less than optimum quality systems may face increased risk of noncompliance and/or regulatory sanctions.

The FDA defines the term "quality system" to include the systems that "assure overall compliance with cGMPs and internal procedures and specifications." Regarding inspections, the FDA uses the term to include the logical individual subsets that may be broken into components that are logically organized within the context of each company's quality control unit and are based on the needs of management with executive responsibility. Under the new systems-based approach, the FDA identified six areas of focus for its inspections:

1. Quality system
2. Facilities and equipment system
3. Materials system
4. Production system
5. Packaging and labeling system
6. Laboratory controls system

Under the quality systems inspection approach, the FDA investigator has the option of conducting an "abbreviated" or "full" inspection. An abbreviated inspection includes the quality system and at least one other system, and a full inspection includes the quality system and at least three other systems. Abbreviated inspections are generally reserved for companies with a satisfactory history of GMP compliance during past two years (e.g., no recalls, no product defects or NDA field alerts, and little shift in its manufacturing profile classes). The full inspection option is intended for companies that deserve a broad/deep evaluation (such as a new firm where the GMP compliance status is unknown) or is in doubt (e.g., history of compliance issues or recidivism).

CBER Risk-Based Inspection Program

On December 1, 2004, the FDA implemented a quality systems–based inspection program for biological drug products (31). CP 7345.848 applies to biological products such as fractionated blood products, allergenic extracts, vaccines, cultured human cells, etc. The program was implemented to provide further guidance to the FDA investigators who had been inspecting biological manufacturers under the Team Biologics initiative that began in 1997 in reaction to contamination issues detected in the plasma fractionation industry. The approach described in CBER's quality system inspection program is substantially the same as for pharmaceuticals (i.e., both involve six quality systems), but the CBER CP contains considerably more details. In "Part V, Regulatory/Administrative Strategy," the CBER version outlines specific evaluation criteria for deciding enforcement actions on the basis of inspection findings for each quality system. The equivalent section of the CDER program is more general and lacks the specificity of the CBER program. While the CBER program is specific to biological drug products, the evaluation criteria have broad applicability to all pharmaceutical drug products, and prudent management will take steps to ensure that these criteria are applied internally when assessing the compliance of quality systems for drug products.

CDRH Risk-Based Inspection Programs

The FDA proposed a revision of the "Medical Device GMP Regulations" on November 23, 1993. After a review of comments received, the agency published the QSR as a final rule on October 7, 1996 under 21 CFR Part 820 (32). In 1999, the FDA issued a manual called Quality Systems Inspection Technique (QSIT) that described a new approach for use during inspections of medical devices (33). The QSR regulations (21 CFR Part 820) contain specific requirements related to quality systems, and although the regulations do not apply to pharmaceuticals, they provide insight into current FDA philosophies, with these concepts having a wider impact than just devices. Under Part 820, the FDA defined the quality system as "the organizational structure, responsibilities, procedures, processes, and resources for implementing quality management" (32). The aim of the quality system is to achieve a state of control through the establishment of conditions and practices that ensure compliance with the intent of sections 501(a) (2)(b) of the act, and the portions of the cGMP regulations that pertain to the quality systems. When quality systems operate under a state of control, companies will produce drug products that have an adequate level of assurance of quality, strength, identity, and purity (34).

EMERGING ISSUES UNDER FDA'S RISK-BASED APPROACH

Since August 2002, the FDA has made a number of important changes in its inspection approach to accommodate risk-based principles. Such changes have been made to focus on high-risk issues, to reduce the time needed to conduct the FDA inspections, and to improve consistency and predictability of the FDA's actions, expectations, and decisions. The agency intends to refocus attention on management's accountability for quality (preventative and corrective actions). The FDA has implemented several significant policy changes and issued a relatively large number of guidance documents that demonstrate that it is serious about allowing companies greater flexibility under the risk-based paradigm.

The FDA has established a group called the Council on Pharmaceutical Quality (Council) that will be responsible for developing policies related to risk management/drug safety, the critical path initiative, and the changes that are related to the FDA initiatives (GMPs for the 21st Century) (35). The Council established a charter to describe the roles and responsibilities of various expert working groups tasked with accomplishing the goals of the Council (36). The Council includes representatives from the CDER, the CBER, the Center for Veterinary Medicine (CVM), the Office of Regulatory Affairs, the Office of the Commissioner, and the Office of the Chief Counsel (OCC). Among the responsibilities of the Council are the following:

- Identify and implement the actions necessary to modernize the regulation of pharmaceutical manufacturing and product quality
- Recognize training needs and opportunities
- Provide oversight of program implementation
- Develop communications
- Provide oversight of international negotiations and expert working groups

The following section describes some of the key changes that have been proposed or implemented by the FDA since launching its GMPs for the 21st Century initiative in August 2002.

Recent FDA Enforcement Trends

The FDA wants to take advantage of advances in scientific and quality improvements by reducing the impediments to improving processes (37). The agency recognizes that, historically, it has been viewed as an impairment to innovation in drug manufacturing. Any changes to a drug's manufacturing process required FDA approval before the change could be implemented. Such strict regulatory control inhibited companies from seeking to apply new technological innovations that could streamline or improve the manufacturing process. With the implementation of quality systems and risk-based focus initiated by the cGMPs for the 21st Century, the emphasis is now on manufacturers demonstrating scientific knowledge of the product itself, rather than focusing primarily on the process by which the product is derived. The FDA will rely on each company's quality system and change control mechanism to monitor changes in manufacturing processes as long as the product's characteristics remain within the defined design space approved by the FDA. Any changes outside this design space would still require FDA approval. This shift in paradigm allows for flexibility in manufacturing while recognizing the company's product knowledge, and allows FDA focus to remain on maintaining the safety of the drug supply.

The FDA hopes to encourage a cooperative relationship with manufacturers, whereby learning and innovation goes both ways. The agency has changed its inspection process from being less focused on punishment to focusing more on the benefits of learning. The FDA hopes to encourage learning by publishing inspection results to both its employees and the industry at large. The goal is to garner consistency across the inspection process by letting industry know the issues that are currently considered important by the FDA. This inspection transparency is intended to encourage compliance rather than punishing noncompliance. The FDA is now encouraging manufacturers to proactively modify their own processes. In 2002, the FDA adopted a revision to its procedures used for issuing warning letters. Previously, the local district could issue warning letters without approval by headquarters. However, since 2002, the OCC must approve all warning letters, and since 2003, the FDA adopted a process requiring approval by the CDER and the CBER before warning letters are forwarded to the OCC for approval. The overall impact of these changes has been a marked decline in warning letters issued by the FDA. Indeed, the number of drug warning letters issued per year has declined from 87 in 2001, when the initiative was first implemented, to a low of 20 per year in 2006 (38).

The FDA is implementing its risk-based model for inspectional oversight by prioritizing manufacturing sites for inspections. The agency uses a risk-based model to assist in predicting where inspections are likely to achieve the greatest public health impact. Manufacturing firms that demonstrate good process and product knowledge are likely to have fewer, less intensive inspections than new companies or those with a history of regulatory non-compliance. By focusing its limited resources on those manufacturing sites requiring the most assistance in regulatory oversight, the FDA intends to maintain the greatest public safety.

Dispute Resolution

Management from a number of companies expressed concerns that they did not have adequate opportunities to raise disputes about GMP or scientific issues that arise during FDA inspections or review of applications. While the FDA did, in fact, have procedures in place for dealing with disputes, it decided in August 2003 to implement a pilot program to formalize the way it deals with disputes for scientific and technical matters related to GMP issues that arise during FDA inspections (general GMPs or PAIs), or during reviews of applications where scientific or technical issues are raised. The one-year program involved a two-tier approach beginning at the local FDA district office and ending with an appeal process, if needed, by a dispute resolution panel at the FDA headquarters (39). The program encouraged discussions of issues during exit meetings when FDA-483 forms were issued, and elevation to the district office if a resolution was not reached during the inspection. Key provisions of this two-tier resolution process are explained below.

Tier One

Disputes are to be filed within 10 days at the local district office (exceptions—international firms are to notify the Division of Field Investigations in the ORA and Team Biologics are to notify the Office of Enforcement in the ORA). When the Tier One office agrees with the dispute, a written response is supposed to follow within 30 days, but if the Tier One review results in disagreement, then it is forwarded to the Center for review.

Tier Two

At the Center, a dispute resolution panel, which includes technical experts from affected centers, performs the second-level review. The membership of the dispute resolution panel excludes agency personnel who made the original decision that is under dispute. Tier Two disputes should be filed within 60 days, and the panel is supposed to notify (in writing) the affected center about their decision.

It is interesting to note that the FDA received only one dispute during the one-year pilot. Tetzlaff published a review article describing the existing procedures for dispute resolution and an outline of the key elements of the draft guidance document that describes the process for dispute resolution (40).

On April 11, 2005, the FDA published "Guidance for Industry and FDA Staff: Submission and Resolution of Formal Disputes Regarding the Timeliness of Premarket Review of a Combination Product" (41). In addition, the agency modified the language on form FDA-483 to include the following statement:

> This document lists observations made . . . during the inspection of your facility. They are inspectional observations, and *do not represent a final Agency determination* regarding your compliance. *If you have an objection* regarding an observation, or have implemented, or plan to implement, corrective action in response to an observation, you may discuss the objection or action with the FDA representative(s) during the inspection or *submit this information to FDA* at the address above. If you have any questions, please contact FDA at the phone number and address above. [emphasis added]

Process Analytical Technology

On September 29, 2004, the FDA published a final guidance to address issues associated with applying a new and novel approach for adopting state-of-the-art technology to enhance assurance of drug safety and quality. "Guidance for Industry PAT—A Framework for Innovative Pharmaceutical Development, Manufacturing, and Quality Assurance" was meant to encourage the introduction and implementation of innovative technologies into the manufacturing process (42). These technologies provide information on materials to improve process understanding and to measure, control, and/or predict quality and performance which may lead to improved efficiency and effectiveness in manufacturing process design and control and quality assurance. These gains in quality and efficiency are dependent on the process and are likely to come from

- Reducing production cycle times by using on-, in-, and/or at-line measurements and controls;
- Preventing rejects, scraps, and reprocessing;
- Effecting real-time release;
- Increasing automation to improve operator safety and reduce human errors;
- Improving energy and material use and increasing capacity; and
- Facilitating continuous processing to improve efficiency and manage variability.

PAT brings a systems perspective to the design and control of manufacturing processes. This integrated quality system orientation allows for a flexible approach to implementing the PAT. For example, the PAT can be implemented using the manufacturer's own quality system. Implementing the PAT enables submission of a comparability protocol to the agency outlining research, validation, and implementation strategies of a proposed manufacturing change, thereby streamlining process innovation.

Comparability Protocols

In September 2003, the FDA published a draft guidance to address the use of comparability protocols for assessing chemistry, manufacturing, and controls (CMC) changes for certain drugs including NMEs, well-characterized proteins, and biologicals. "Comparability Protocols Protein Drug Products and Biological Products—Chemistry, Manufacturing, and Controls Information" describes recommendations for preparing and using predefined change evaluation plans known as comparability protocols (43). A comparability protocol is a comprehensive, detailed plan that describes the specific tests and studies, analytical procedures, and acceptance criteria to be achieved to demonstrate the lack of adverse effect for a specified type of CMC change that may relate to the safety or effectiveness of the drug product (44). A comparability protocol demonstrates that a manufacturer has sufficient process understanding to implement a change, and may allow for the manufacturer to implement a CMC change without obtaining prior approval from the FDA, thereby allowing for a product to be put into distribution sooner than if the protocol were not used. In some cases, a comparability protocol may preclude supply disruption or shortages.

Electronic Records and Signatures (21 CFR Part 11)

The FDA's goal in issuing Part 11 was to provide criteria under which the "agency considers electronic records, electronic signatures, and handwritten signatures executed to electronic records, to be trustworthy, reliable, and generally equivalent to paper records and handwritten signatures executed on paper" (44). Part 11 provides recommendations about computer systems that are used to create, modify, maintain, archive, retrieve, or transmit clinical data for submission to the FDA. The agency bases its decisions on this data, so the data must be guaranteed to be of high quality and integrity.

While Part 11 was written with FDA requirements in mind, it duplicated other regulations already in place and left open the potential that the industry would have to implement additional computer systems or maintain old ones in order to meet the specific requirements of Part 11, all without gaining any actual technological advantage. In addition, since Part 11 regulations were issued, Congress has enacted the Government Paperwork Elimination Act (GPEA) (45), which requires federal agencies to accept electronic records and signatures in satisfaction of programmatic requirements. Under the GPEA, the implementation of electronic record systems can be pursued under existing regulatory requirements, and does not require the issuance of separate regulatory schemes such as Part 11.

This redundancy and potential for added expense led the industry to form a coalition comprising of 13 trade associations representing manufacturers of FDA-regulated products including foods, drugs, cosmetics, veterinary drugs, and medical devices. The Industry Coalition on 21 CFR Part 11 filed a citizens' petition requesting the FDA abolish Part 11 in its entirety (46), stating, "These aspects of Part 11 are not only burdensome to industry, but have also made enforcement of electronic systems security needlessly burdensome for FDA."

The FDA withdrew previous draft guidance documents and reissued a new guidance that provided a certain degree of regulatory relief to the industry. The agency announced its intention to use enforcement discretion for certain Part 11 sections that related to validation and record keeping (47). Current requirements regarding electronic records are set forth in FDA rules (other than Part 11) and are known as predicate rules. "Predicate rules" have been characterized by the FDA as "the underlying requirements set forth in the [Federal Food, Drug, and Cosmetic] Act, the PHS Act, and FDA regulations (other than Part 11)" (47). These are rules the FDA will strictly enforce while applying enforcement discretion in interpreting Part 11 requirements.

While comments and the citizens' petition are under review by the agency, Part 11 remains in effect, as the FDA reviews provisions in that regulation.

Computerized Systems in Clinical Trials

The computerized systems used to support clinical trials are covered under 21 CFR Part 11. To better explain the FDA's interpretation of Part 11 with regard to clinical trials while Part 11 is being reviewed, draft guidance for industry, "Computerized Systems Used in Clinical Trials" (48), was issued in September 2004. This guidance explains requirements for the computerized system data entry, features, security, dependability, and controls necessary to guarantee data quality, accuracy, integrity, and reliability.

Aseptic Processing Guidance

On September 28, 2004, the FDA published the final guidance that describes risk-based approaches for aseptic processing "Sterile Drug Products Produced by Aseptic Processing—Current Good Manufacturing Process" (49). Sterile drug products are an important part of FDA's risk-based inspection program because of their therapeutic significance. The focus of these inspections is on prevention rather than removal of contamination, so adequate design and planning of aseptic processing facilities are a must. Many manufacturers are utilizing automation and isolation processes to protect exposed sterile drug products during aseptic manufacture.

This guidance advocates a risk-based and quality system framework that emphasizes contamination prevention. It also delineates the roles of personnel, design, environmental controls, and media fills on aseptic drug production. The goal of this guidance is to encourage contamination prevention, avoidable risks reduction, and thereby helping to ensure the availability of these therapeutically significant pharmaceuticals.

Validation of Conformance Batches

One recent change illustrates the FDA's recognition of the role of emerging technologies that may be used for process validation. Since the issuance of the 1987 guideline of process validation (50), the FDA position on process validation remained unchanged, and for more than a decade, it was common practice to validate new or modified processes by producing three batches at full commercial sizes. On March 12, 2004, the FDA issued a new Compliance Policy Guide (CPG) that establishes a new policy with respect to requirements for conformance batches (also called three-lot validation) (51). The agency eliminated the requirement for production of multiple conformance batches for finished pharmaceuticals and APIs but has maintained that production processes must be validated. With the implementation of this new policy, the NDAs may be approved before a firm has manufactured "conformance batches," and the FDA now permits (under certain conditions) drugs to be marketed while data are being gathered to confirm the validity of the manufacturing process. This policy covers products regulated by the CDER, the CBER, and the CVM, but does not apply to biologics covered by the BLA or recombinant proteins covered by the NDAs. While the FDA has eliminated its reference to "three" validation batches, it stresses the importance of post-marketing information.

This change reflects the agency's willingness to allow companies that apply risk-based approaches that increase process understanding to have an opportunity to apply alternate approaches for validation. The new policy clarifies the FDA expectations for the timing of the completion of process validation relative to the filing of new drug applications. The FDA has not eliminated the requirement to validate processes, but provides flexibility to companies that apply sound scientific approaches. If alternative approaches are used, the agency will expect companies to have documented justifications and supporting data to support the approach(es) used. The FDA now allows companies an opportunity to gain approval of the NDAs before they have completed the conformance batches portion of their validation programs, but it has not changed its requirement to complete such validation prior to commercial distribution of the product.

The following are a few of the key changes that are outlined in the CPG. During the PAIs, the FDA will audit all available information about process validation whether or not validation has been completed. The agency will review protocols, test results, and final reports. It will focus on completed validation reports. If the records show the process is not under control or data are of questionable integrity (and the company has not committed to making appropriate changes), the district office will likely recommend withholding approval of the pending application. If the deficiencies found by the FDA also apply to approved product(s) produced by a process similar to the PAI product, then the agency will withhold approval of the pending application, and may issue a warning letter or invoke regulatory sanction.

Transfer of Therapeutic Biologics from CBER to CDER

On June 26, 2003, the FDA announced that therapeutic biologicals would no longer be the responsibility of the CBER, and the BLAs would be transferred to the CDER including the following product classes (52):

* In vivo monoclonal antibodies
* Cytokines, growth factors, enzymes, immunomodulators, and thrombolytics
* Therapeutic proteins extracted from animals or microorganisms (including recombinant products and excluding clotting factors)
* Other non-vaccine therapeutic immunotherapies

The CDER will retain responsibility for the following product classes (except those retained by the CBER):

* Blood, blood components, and related products
* In vitro diagnostics
* Viral-vectored gene insertions (gene therapy)
* Products from human or animal cells
* Allergenics and allergen patch tests
* Antitoxins, antivenins, and venoms
* Vaccines (including therapeutic vaccines)
* Toxoids and toxins intended for immunization

The transfer to the CDER was effective on June 30, 2003 and included all affected BLAs as well as relocation of FDA personnel from the CBER to the CDER. The CDER is now responsible for pre-market review of the BLAs for therapeutic biologics. The FDA maintains a listing of products transferred and those that remained with the CBER (52).

On October 1, 2003, the CBER's Office of Therapeutics Research and Review was transferred into two newly created offices within the CDER (i.e., Office of Drug Evaluation VI and Office of Biotechnology Products) (53). These two offices will be responsible for the following:

- Monoclonal antibodies for in vivo use
- Cytokines, growth factors, enzymes, immunomodulators, and thrombolytics
- Proteins intended for therapeutic use that are extracted from animals or microorganisms, including recombinant versions of these products (except clotting factors)
- Other non-vaccine therapeutic immunotherapies

On March 24, 2005, the FDA published in the *Federal Register* the new organization and contact information (54).

Phase 1 Drugs Exempted from Certain cGMP Requirements

On January 12, 2006, the FDA issued a draft guidance, "INDs—Approaches to Complying with CGMP During Phase 1," detailing requirements for producing drug and biological products for Phase 1 development (55). Manufacturers of clinical trial materials are required to meet the requirements of Part 211 (21 CFR 211) in order to comply with § 501(a)(2)(B) of the FD&C Act. Part 211 treats all drug products the same regardless of development phase, requiring producers of clinical trial materials to meet commercial-scale cGMP requirements (56). The new guidance was designed to reduce the financial burden placed on industry by Part 211 by exempting Phase 1 drugs from complying with the requirements in the FDA regulations. The FDA believed that using appropriate manufacturing controls during each stage of drug development would not only ensure subject safety (the main purpose of Phase 1 studies) but also product quality.

On the basis of significant adverse comments received during the comment phase, the FDA withdrew this draft guidance on May 2, 2006 (57).

Pharmaceutical Inspectorate

On February 27, 2003, the FDA established the Pharmaceutical Inspectorate (PI) to aid in pre-approval and cGMP inspections (58). The FDA recognized the value a team approach offered to both inspectors and the industry by ensuring inspection teams consist of members knowledgeable in both the science and technology of manufacturing as well as regulatory requirements.

The PI is a cadre of experts specially trained in inspection techniques for facilities making complex or high-risk drugs. In addition to their regulatory expertise, members of the PI are also trained to be experts in the science and technology of pharmaceutical manufacturing. The FDA has developed an entire training curriculum for members of the PI, including modules on regulating pharmaceutical quality and the relationship to the FDA's mission, risk management, advanced quality systems, pharmaceutical science, current regulatory programs and procedures, technology, and investigations. By continually updating training, the FDA ensures investigators are current on all advances in pharmaceutical production enabling them to make knowledgeable decisions during inspections.

FDA's International Collaborations

With the advent of a global economy, the FDA has sought to harmonize its quality and scientific standards and requirements with other international health and regulatory agencies. The agency is collaborating with such forums as the ICH, the European Medicines Agency (EMEA), the Ministry of Health, Labor and Welfare, Japan (MHLW), and the World Health Organization (WHO) (59). In November 2003, the ICH began work on developing an internationally harmonized plan for a pharmaceutical quality system on the basis of an integrated approach to science and risk management (2).

The ICH is working with representatives from regulatory agencies in the U.S.A., the European Union, and Japan to make the international drug regulatory process more efficient and uniform. This work will help make new drugs available with minimum delays to both American consumers and those in other countries (60). One outcome of this collaboration is the common technical document (CTD), which is a standard format for submitting safety and efficacy information about a new drug applicable internationally (61). The CTD simplifies the drug approval process by unifying the requirements of each international regulatory agency into a single format.

In addition, the FDA has announced its intentions to join the Pharmaceutical Inspection Cooperation Scheme (PIC/S) in an effort to promote harmonized GMPs and quality systems on a worldwide basis (2). Membership in the PIC/S will allow the FDA to build working relationships with international authorities enabling mutual confidence and mutual training of inspectors.

The FDA, in cooperation with the ICH, is a member of three working groups that are developing a harmonized approach to drug safety and risk management. New ICH initiatives such as Q8—Pharmaceutical Development, Q9—Quality Risk Management, and Q10—Pharmaceutical System, demonstrate how it intends to fulfill its mission to facilitate the dissemination and communication of information on harmonized guidelines and their use in encouraging the implementation and integration of common standards (62).

SUMMARY AND CONCLUSIONS

This chapter reviewed some of the key changes made by the FDA since it announced its risk-based approach to pharmaceutical GMPs for the 21st Century in August 2002 (1,13,63). The FDA initiative provides the agency with a way to reduce the resources needed to monitor the industry for GMP compliance, and the FDA is changing its inspection approach and enforcement posture (i.e., the FDA has been issuing fewer warning letters and certain regulatory sanctions have declined).

The risk-based approach to pharmaceutical GMPs is changing the way FDA is regulating the pharmaceutical industry, and the agency is encouraging companies to apply some new and novel approaches for quality management. In the past, management of pharmaceutical manufacturing facilities may have avoided some innovative changes and/or improvements because of uncertainties about gaining timely approvals of the FDA supplements and/or the cost burdens associated with validating new applications.

The FDA's risk-based approach to quality management opens the door for companies to apply some new and innovative approaches. Now the industry has an unprecedented opportunity to introduce improvements to its quality management programs to help increase assurances of product quality and safety, while at the same time reducing costs and the time it takes to get products to the market. The ultimate objective of the FDA initiative is to encourage companies to apply sound scientific principles and risk management tools to better manage the hazards during manufacturing of pharmaceuticals. When risk-based approaches are successfully adopted, companies will ultimately be able to eliminate manufacturing controls that do not add value to ensuring drug safety or quality.

By applying effective risk-based approaches during the drug development lifecycle (i.e., from discovery though post-approval changes), management in the pharmaceutical industry have a unique window of opportunity to optimize manufacturing processes, reduce time to market, and prevent and detect regulatory compliance issues. Prudent management will move quickly to leverage this opportunity by establishing a risk-based approach to managing quality while FDA is embracing such changes. This chapter outlines a number of key changes and the impact of the FDA's new approach to inspections under its quality systems paradigm and during its risk-based review of the NDAs and BLAs.

It seems likely that at some point in time the CDER's Office of New Drug Chemistry will eventually modify the filing requirements for the CMC section to apply a new risk-based model for the content and review criteria. It is impossible to predict the content of a new system and its implementation date, but it seems clear that the agency and the industry would benefit from applying risk-based approaches to the CMC data. The FDA has already issued a number of guidance documents related to 21 CFR Part 314.70 to reduce the burden of manufacturing supplements (i.e., to integrate CMC reviews and inspections). It intends to allow companies more flexibility to make changes without the need for prior approvals and will use post-approval inspections to audit the data and information in support of changes (including validation).

A brief review of the FDA enforcement practices was offered, because a number of companies continue to experience difficulties during the FDA inspections. While regulatory sanctions are generally at levels lower than a decade ago, there are notable exceptions, and a number of companies have been "surprised" during violative FDA inspections of facilities that had no prior warning of serious compliance issues. A number of enforcement trends were described, and the FDA is more actively collaborating with other government agencies (such as the OIG and the SEC). Under the FDA's risk-based approach, companies can significantly improve the odds of a successful inspection by using systematic techniques to prepare before the FDA arrives and by making certain personnel know what to expect. By all indications, it appears that the FDA will continue its risk-based approach to decide which companies to inspect, how often, and to what degree of coverage. Companies that have adopted modern quality systems to gain a better understanding of their processes and sources of variability will tend to have successful outcomes during the FDA inspections. In contrast, companies that fail to heed the FDA's warning signals and who have not availed themselves of this once-in-a-lifetime opportunity to adopt risk management principles are likely to face increased regulatory scrutiny and/or regulatory sanctions. Hopefully, the information presented in this chapter will be of value to those who strive to understand the FDA expectations.

REFERENCES

1. FDA. FDA unveils new initiative to enhance pharmaceutical good manufacturing practices [press release]. August 21, 2002; P02–P28. Available at: http://www.fda.gov/bbs/topics/NEWS/2002/NEW00829.html.
2. FDA. Pharmaceutical cGMPs for the 21st Century: A Risk-Based Approach Final Report—Fall 2004. Available at: http://www.fda.gov/cder/gmp/gmp2004/GMP_finalreport2004.htm. Accessed October 11, 2006.
3. FDA. Challenge and Opportunity on the Critical Path to New Medical Products: Innovation or Stagnation? Figure 2, March 2004. Available at: http://www.fda.gov/oc/initiatives/criticalpath/whitepaper.html.
4. FDA. CDER Report to the Nation: 2002. Available at: http://www.fda.gov/cder/reports/rtn/2002/rtn2002.htm#Contents.
5. FDA. Prescription Drug User Fee Act. September 1992. Available at: http://www.fda.gov/oc/pdufa/default.htm.
6. Pharmaceutical Research and Manufacturers of America. Pharmaceutical Industry Profile 2005. Washington, DC: PhRMA.
7. FDA. Challenge and Opportunity on the Critical Path to New Medical Products. March 2004. Available at: http://www.fda.gov/oc/initiatives/criticalpath/whitepaper.html.
8. Pazdur R. Statement of before the subcommittee on criminal justice, drug policy and human resources committee on government reform, U.S. House of Representatives. September 7, 2005. Available at: http://www.fda.gov/ola/2005/womenandcancer 0908/attachmentC.html.
9. Abboud L, Hensley S. Factory shift, new prescription for drug makers: update the plants. Wall Street Journal. September 3, 2003:293(48).
10. FDA. FDA Letter from Deputy Commissioner to Employees: Agency Wide, Pharmaceutical cGMPs Initiative and Pharmaceutical cGMP Initiative Questions and Answers. August 21, 2002. Available at: http://www.fda.gov/oc/guidance/qsas.html.

11. FDA. Current Good Manufacturing Practice; Amendment of Certain Requirements for Finished Pharmaceuticals; Withdrawal. Federal Register, December 4, 2007; 72(232):68111–68113. Available at: http://a257.g.akamaitech.net/7/257/2422/01jan20071800/edocket.access.gpo.gov/2007/pdf/E7-23271.pdf.

12. FDA. Amendment to the Current Good Manufacturing Practice Regulations for Finished Pharmaceuticals. Federal Register, December 4, 2007; 72(232):68064–68070. Available at: http://a257.g.akamaitech.net/7/257/2422/01jan20071800/edocket.access.gpo.gov/2007/pdf/E7-23294.pdf.

13. FDA. Advancing America's health. Advancing medical breakthroughs [press release]. March 16, 2004; P04–30. Available at: http://www.fda.gov/bbs/topics/news/2004/NEW01035.html and complete report at: http://www.fda.gov/oc/initiatives/criticalpath/whitepaper.pdf.

14. FDA. The critical path: accelerating the development of medical products. FDA Consumer Magazine, September/October 2004. Available at: http://www.fda.gov/oc/speeches/2004/nma0804.html.

15. FDA. FDA unveils critical path opportunities list outlining blueprint to modernizing medical product development by 2010 [press release]. March 16, 2006; P06–39.

16. FDA. Guidance for Industry:Premarketing Risk Assessment. March 2005. Available at: http://www.fda.gov/cder/guidance/6357fnl.htm.

17. FDA. Guidance for Industry: Development and Use of Risk Minimization Action Plans. Draft Guidance, March 2005. Available at: http://www.fda.gov/cder/guidance/6358fnl.htm.

18. FDA. Guidance for Industry: Good Pharmacovigilance Practices and Pharmacoepidemiologic Assessment. Draft Guidance, March 2005. Available at: http://www.fda.gov/cder/guidance/6359OCC.htm.

19. FDA. "Drug Watch" for Emerging Drug Safety Information. Draft Guidance, May 2005. Available at: http://www.fda.gov/cder/guidance/6657dft.htm.

20. FDA. FDA Publishes Guidance on Communication of Drug Safety Information. March 2, 2007. Available at: http://www.fda.gov/bbs/topics/NEWS/2007/NEW 01577.html.

21. FDA. 21 CFR Parts 210 and 211. Current Good Manufacturing Practices for Finished Pharmaceuticals. Federal Register, September 28, 1978; 43(190):45013–45336.[docket no. 73N-0339]. Available at: http://www.fda.gov/cder/dmpq/preamble.txt.

22. FDA. Current Good Manufacturing Practice: Proposed Amendment of Certain Requirements for Finished Pharmaceuticals[proposed rule] 61. Federal Register, May 3, 1996; 61(87):20103.

23. FDA. Guidance: Quality Systems Approach to Pharmaceutical Current Good Manufacturing Practice Regulations. September 2006. Available at: http://www.fda.gov/ohrms/dockets/ac/05/briefing/2005-4136b1_05_pharmaceutical%20CGMP.pdf.

24. ISO. ISO-9000. Available at: http://www.iso.org/iso/iso_catalogue/management_standards/iso_9000_iso_14000.htm.

25. ICH. Guidance for Industry: Q8 Pharmaceutical Development. May 2006; Guidance for Industry: Q9 Quality Risk Management. June 2006; and Draft Consensus Guideline: Pharmaceutical Quality System Q10. May, 2007.

26. FDA. 21 CFR Part 820. Quality Systems Regulations. Available at: http://www.accessdata.fda.gov/scripts/cdrh/cfdocs/cfcfr/CFRSearch.cfm?CFRPart=820&show FR=1.

27. Tetzlaff RF, Shepherd RE, Leblanc AJ. The validation story: perspectives on the systematic GMP inspection approach and validation development. Pharm Technol 1993; 17(3):100–116.
28. FDA. Pre-approval Inspection Program. Compliance Program Guidance Manual CP 7346.832. Revisions dated September 3, 2003, and March 2004. Available at: http://www.fda.gov/cder/gmp/PAI-7346832.pdf.
29. FDA. Pre-approval Inspections/Investigations. Compliance Program Guidance Manual CP7356.832. Chapter 46. September 2003. Available at: http://www.fda.gov/cder/gmp/PAI-7346832.pdf.
30. FDA. Drug Manufacturing Inspections. Compliance Program Guidance Manual CP7356.002. Available at: http://www.fda.gov/cder/dmpq/compliance_guide.htm; http://www.fda.gov/ora/cpgm/7356_002/7356-002FINAL.pdf.
31. FDA. Inspection of Biological Drug Products. Compliance Program Guidance Manual CBER 7345.848, December 1, 2004. Available at: http://www.fda.gov/cber/cpg/7345848.htm.
32. FDA. Medical Devices; Current Good Manufacturing Practice (cGMP) Final Rule; Quality System Regulation, 21 CFR Part 820. Federal Register, October 7, 1996; 61 (195):52602–52662.
33. FDA. Quality Systems Inspection Technique. August 1999. .Available at: http://www.fda.gov/ora/inspect_ref/igs/qsit/qsitguide.htm.
34. ISO. 8402:1994 [Note: Superseded by ISO 9000:2000]. Available at: http://www.iso.org/iso/iso_catalogue/catalogue_tc/catalogue_detail.htm?csnumber=20115.
35. FDA. Charter of the Council on Pharmaceutical Quality. Available at: http://www.fda.gov/cder/gmp/gmp2004/charter.htm.
36. FDA. Staff Manual Guides, Volume III—General Administration, FDA Official Councils and Committees. SMG 2010.5—Charter of the Council on Pharmaceutical Quality. Available at: http://www.fda.gov/smg/vol3/2000/2010_5.html.
37. FDA. Risk-Based Method for Prioritizing cGMP Inspections of Pharmaceutical Manufacturing Sites: A Pilot Risk Ranking Model. September 29, 2004. Available at: http://www.fda.gov/cder/gmp/gmp2004/risk_based_method.htm.
38. F-D-C Reports. The Gold Sheet. April 2007; 41(4). 'Drug GMP Warning Letters for FY 2006'.
39. FDA. Guidance for Industry: Formal Dispute Resolution—Scientific and Technical Issues Related to Pharmaceutical cGMP. August 2003. Available at: http://www.fda.gov/cder/guidance/5804dft.htm.
40. Tetzlaff RF. Resolving scientific and technical disputes. BioPharm Int S-28, November 2003.
41. FDA. Guidance for Industry and FDA Staff: Submission and Resolution of Formal Disputes Regarding the Timeliness of Premarket Review of a Combination Product. April 11, 2005. Available at: http://www.fda.gov/cber/guidelines.htm#combo.
42. FDA. Guidance for Industry PAT: A Framework for Innovative Pharmaceutical Development, Manufacturing, and Quality Assurance. September 29, 2004. Available at: http://www.fda.gov/cder/guidance/6419fnl.htm.
43. FDA. Comparability Protocols Protein Drug Products and Biological Products: Chemistry, Manufacturing, and Controls Information. September 2003. Available at: http://www.fda.gov/cder/guidance/protcmc.pdf.
44. FDA. Electronic Records: Electronic Signatures Final Rule, 21 CFR Part 11. Federal Register, March 2000. Available at: http://www.fda.gov/ora/compliance_ref/part11/frs/background/11cfr-fr.htm.

45. Government Paperwork Elimination Act (GPEA). Pub. L. No. 105-277, Div. C., Tit. XVII, §§ 1701 *et seq*. October 21, 1998.

46. Citizens Petition. Industry Coalition on 21 CFR Part 11. September 17, 2004. Available at: http://www.fda.gov/ohrms/dockets/dailys/04/sep04/092004/04p-0429-cp00001-vol1.pdf.

47. FDA. Guidance for Industry: Part 11, Electronic Records, Electronic Signatures— Scope and Application. September 3, 2003. Available at: http://www.fda.gov/cder/guidance/5667fnl.htm.

48. FDA. Computerized Systems Used in Clinical Trials. September 28, 2004. Available at: http://www.fda.gov/cder/guidance/6032dft.htm.

49. FDA. Sterile Drug Products Produced by Aseptic Processing: Current Good Manufacturing Process. September 28, 2004. Available at: http://www.fda.gov/cder/guidance/5882fnl.htm.

50. FDA. Guideline of General Principles of Process Validation. May 1987. Available at: http://www.fda.gov/cder/guidance/pv.htm.

51. FDA. Process Validation Requirements for Drug Products and Active Pharmaceutical Ingredients Subject to Pre-Market Approval, CPG 7132c.08. March 12, 2004. Available at: http://www.fda.gov/ora/compliance_ref/cpg/cpgdrg/cpg490-100.html.

52. FDA. Transfer of Therapeutic Products to the Center for Drug Evaluation and Research. Letter to Industry. June 26, 2003. Available at: http://www.fda.gov/cber/transfer/transfer.htm or CDER Office of Communication, Training and Manufacturers Assistance.

53. FDA. Therapeutic Biological Products. April 9, 2007. Available at: http://www.fda.gov/cder/biologics/default.htm.

54. FDA. Food and Drug Administration; Drug and Biological Product Consolidation; Addresses; Technical Amendment. Federal Register, March 24, 2005; 70:14978–14986.

55. FDA. Guidance for Industry INDs: Approaches to Complying with cGMP During Phase 1. Draft Guidance, January 2006. Available at: http://www.fda.gov/cder/ guidance/6164dft.htm.

56. FDA. Guideline on the Preparation of Investigational New Drug Products (Human and Animal). March 1991. Available at: http://www.fda.gov/cder/guidance/old042fn.pdf.

57. FDA. Current Good Manufacturing Practice Regulation and Investigational New Drugs; Withdrawal. Federal Register, May 2, 2006; 71:25747. Available at: http://www.fda.gov/ohrms/dockets/98fr/06-4091.htm.

58. FDA. Progress Report of the Pharmaceutical Inspectorate Working Group. February 27, 2003. Available at: http://www.fda.gov/cder/gmp/pharminspectorate.htm.

59. FDA. CDER Report to the Nation: 2005. August 14, 2006. Available at: http://www.fda.gov/cder/reports/rtn/2005/rtn2005-5.HTM#Agreements.

60. FDA. International Activities. Available at: http://www.fda.gov/cder/audiences/iact/iachome.htm#ICH.

61. ICH. M4: Common Technical Document. August 11, 1999. Available at: http://www.fda.gov/cber/gdlns/ichm4.pdf.

62. ICH. Revised ICH Terms of Reference. Available at: http://www.ich.org/cache/html/581-272-1.html.

63. FDA. A Science and Risk-Based Approach to Product Quality Regulation Incorporating an Integrated Quality Systems Approach. August 21, 2002. Available at: http://www.fda.gov/oc/guidance/gmp.html.

3

Critical Role of the Pharmaceutical Scientist in Product Development and Preparing for Pre-Approval Inspections

Mahdi Fawzi, Richard Saunders, and Parimal Desai
Wyeth Research, Pearl River, New York, U.S.A.

INTRODUCTION

The underlying message in the Food and Drug Administration (FDA) guidance document "Pharmaceutical cGMPs for the 21st Century: A Risk Based Approach" (1), issued in 2002, states that regulatory review and inspection should be based on state-of-the-art science and also encourages adaptation of new technologies by scientists.

In line with the above expectation, the role of the pharmaceutical scientist in drug development has been changing over the last two decades as considerable progress and understanding have occurred in the areas of pharmacokinetics and pharmacodynamics. Conventional formulation development activities have progressively changed to more targeted drug delivery systems. While typical dosage forms in the eighties were tablet, capsules, injectables, semisolids, and suppositories, there have emerged newer delivery systems such as drug-eluting stents, transdermal patches, needleless injectables, and depot therapies. Once-a-day regimens in oral drug administration have become more a standard than the exception in dosing.

In the past, much emphasis was placed on formulation development; however, the emphasis on the manufacturing technology area did not get the same attention. There was no systematic approach in the industry for technology transfer to production sites. The result was repeated failures of manufacturing processes due to a variety of reasons that became summarized as a "Non-Robust Process." The FDA and industry leaders met together to understand the overall situation, which gave rise to new initiatives, such as Pharmaceutical cGMPs for the 21st Century: A Risk Based Approach (1), Quality by Design (QbD) (2), and Process Analytical Technology (PAT) Guidelines (3), aimed to jump-start the innovative process in dosage form design and manufacturing sciences.

As a potential new drug candidate moves into development, more formalized quality standards are employed—Good Laboratory Practice (GLP), Good Clinical Practice (GCP), and Good Manufacturing Practice (GMP). In addition to these comprehensive rules, scientists need to ensure that they understand all the characteristics of the new drug—solubility, stability, physical form, interaction with other substances, interaction with receptors (both target and non-target receptors), toxicology, pharmacology, etc. Similarly, the chemists and formulators need to be confident that they know how to reliably make the drug substance and the formulations every time and that they understand the critical features of the manufacturing processes to ensure the continued production of a consistent drug product.

The pharmaceutical scientist has been challenged in a very rapid move to develop not only robust formulations but formulations that can be manufactured economically and safely on the basis of sound scientific principles. Pharmaceutical scientists are taking a fresh look at understanding underlying mechanistic-based scientific approaches in formulation development and manufacturing development. The industry is starting to understand the benefits of efforts such as regulatory flexibility in making further changes and thus saving time and money.

Wyeth has been at the forefront of such initiatives and has taken a leadership role by participating in the FDA pilot program by experimenting with QbD and PAT approaches in drug development. Considerable emphasis has been placed on building scientific considerations right from the early drug development phases. In this chapter, you will read examples of such novel approaches in dosage form development from the early development phase through to the commercialization phase, including the pre-approval inspection (PAI) process. We will start with various approaches taken at Wyeth Research and Development, which are covered in eight sections:

1. Formulation Development from Predevelopment to Phase 1 Studies
2. In Vitro/In Vivo Correlation in Drug Development
3. Conventional Oral Drug Delivery
4. Controlled Drug Delivery
5. Novel Drug Delivery Systems
6. Parenteral Drug Development Challenges
7. The Role of Packaging Science in Drug Development
8. Application of Process Analytical Technology in Early Development

FORMULATION DEVELOPMENT FROM PREDEVELOPMENT TO PHASE 1 STUDIES

The pharmaceutical scientist in early development has the critical role of selecting the right active pharmaceutical ingredient (API) solid form and determining the right delivery system for first-in-human (FIH) dosing, often with very limited compound availability and compressed timelines. The decisions that are made at this early developmental stage for the selection of salt/polymorph and dosage form can significantly impact the product development portfolio for PAIs and the quality of chemistry, manufacturing, and control (CMC) dossier for ultimate registration.

Discovery Lead Selection

By interacting with the discovery function or organization even before candidate selection, the pharmaceutical scientist can delineate the chemical and bio-pharmaceutical limitations of the drug substance and facilitate lead optimization for the best candidates for FIH. Pharmaceutical profiling is typically done to select the compound that not only has the desired drug activity, but also the physical, chemical, and biopharmaceutical properties that will translate into bioavailability in humans.

Typically, a number of in silico chemical structure–based calculations are used in conjunction with in vitro pharmaceutical profiling, followed by testing leads in animals to evaluate the prediction in vivo. Many clinical candidates, however, still enter development with less-than-desirable properties for delivery after in vivo testing.

After screening has identified candidate compounds for potential clinical development, a thermodynamically stable, formulatable, and bioavailable crys-talline form must be selected. The crystalline free acid or base is evaluated for crystallinity, hygroscopicity, solubility, purity, and stability. If these properties are unfavorable, or these forms cannot be manufactured with a high degree of polymorphic and chemical purity, then salt selection can be pursued. Should a change of salt form be necessary later in the cycle, significant preclinical studies would need to be repeated unless biopharmaceutical equivalency can be demonstrated across the two forms. Since toxicology dosing is often done with suspensions, the potential for conversion of the drug substance into a different polymorph or hydrate must be understood. Use of the thermodynamically favored polymorph will help prevent this conversion.

Drug Safety Formulations

The pharmaceutical scientist plays an important role in the selection of the vehicle for preclinical drug safety (toxicology) studies. If the physical and chemical properties of the drug substance are favorable, the preferred vehicle is a neat powder in a capsule for non-rodent species and an aqueous solution or suspension for rodents. The toxicological formulations should achieve animal exposures that allow for high-dose toxicology assessment. If solubility is low, surfactants or solubilizers may be added to enhance solubility. The toxicology formulation should also have adequate stability for repeated dosing of the animals. In some cases, the formulation may need to incorporate a coating for a delayed or sustained release to improve the tolerability of the drug in animals, or to simulate clinical formulations. In cases where the bioavailability of the drug is still too low for toxicology studies, the pharmaceutical scientist may develop more sophisticated formulations with enhanced bioavailability for animal testing. This work may take considerable time and collaboration. All available pharmaceutical property data for the compound must be reviewed, and additional preformulation studies conducted. In some cases nanodispersions, which contain extremely fine drug particles, may provide the needed exposure in animals. In other cases, additional screening of a series of high-energy or solubilized formulations both in vitro and in vivo may lead to adequate exposure. Collaboration with toxicology is necessary to ensure that the dosing vehicles do not produce adverse effects in animals in the quantities used.

Polymorphism and Hydrate Formation

Understanding polymorphism and hydrate formation is critical to the manufacturing and commercial success of a drug product. Polymorphism can affect the product's quality, safety, and efficacy. Pharmaceutical manufacturers have recalled commercial products because of poor dissolution and lack of efficacy when the drug converts to a less soluble polymorph during its shelf life. A polymorphism screening study to identify all potential forms of a drug substance, and the conditions under which changes can occur, should be done as soon as possible after the acid/base or salt form is selected. For a polymorphic drug substance, the FDA has drafted guidelines for investigating and controlling the form in a solid or suspension dosage form. Decision trees have also been developed to guide specifications for drug product testing (4,5). A thorough understanding of the polymorphic behavior of a drug substance in the early stages of development will enable QbD by selecting the optimum form prior to FIH studies and by monitoring and controlling the polymorph as the drug product moves toward commercialization.

FIH Formulation Development

The current pharmaceutical industry trends follow two approaches for FIH formulations: (1) active-only formulations, e.g., powder-in-bottle or powder-in-capsule approach, or (2) traditional formulations. There are advantages and disadvantages with both approaches. While the active-only approach can move quickly into phase 1 to assess drug safety, this formulation is not feasible beyond the initial stage, and additional time and resources are needed to develop a phase 2 formulation that can be manufactured at scale in parallel to the FIH study. In addition, a bridging pharmacokinetic study in humans would be necessary to ensure the consistent safety and efficacy with the new traditional formulation. For compounds with poor solubility and/or poor permeability, bioequivalency between the above two types of formulations may be disparate, and adjustment of dosages may be necessary. A traditional formulation approach can reduce the time and resources between clinical phases and shorten the overall drug development cycle. However, it requires more work upfront where the drug candidates have a higher failure rate in clinical trials.

At Wyeth, the bona fide formulation approach is used to enable building quality into the clinical outcome from the beginning and to reduce R&D cycle time. Risk is minimized through a rational formulation development strategy based on the Biopharmaceutics Classification System (BCS) (6).

Wyeth, like many other global pharmaceutical companies, also coordinates its formulation development efforts to incorporate excipients that are acceptable in all countries. A list of new drug application (NDA)-approved levels of excipients in formulations can be accessed at the Center for Drug Evaluation and Research home page on the FDA's Web site (7). For the introduction of a new excipient, the FDA has established guidelines for safety studies (8).

The BCS is used by the FDA (6,1) as a means to make decisions on the requirement of in vivo bioavailability and bioequivalence testing of immediate-release drug products on the basis of the solubility and permeability criteria. Drugs are categorized into four classes on the basis of solubility and permeability as shown in Table 1. Formulation strategies employed at Wyeth over the past 10 years for the different classes are also given in the table.

Many new compounds do not fall within the BCS class 1 category. Thus, there is increasing interest in the development of lipid and self-microemulsifying

Table 1 BCS and Formulation Strategies

BCS class	Solubility	Permeability	Formulation strategy
1	High	High	• Conventional capsule or tablet
2	Low	High	• Micronized API and surfactant • Nanoparticle technology • Solid dispersion • Melt granulation/extrusion • Liquid or semisolid filled capsule • Coating technology
3	High	Low	• Conventional capsule or tablet • Absorption enhancers
4	Low	Low	• Combination of BCS 2 and absorption enhancers

Abbreviations: API, active pharmaceutical ingredient; BCS, biopharmaceutics classification system.

dosage forms. These liquid and semisolid formulations facilitate the delivery of highly lipophilic or hydrophobic compounds, which have good activity in Discovery screening but poor deliverability. These formulations also require specialized equipment that must either be purchased or outsourced. Early communication of this need will make a smoother transition into the later development stages.

Formulation Stability

Development of the most stable formulation begins in preformulation where degradation mechanisms of the drug substance are elucidated. Moisture, pH, oxygen, trace metals, and light can be accelerators of degradation, and early identification of the degradants from accelerated studies will provide invaluable information for formulation development. Excipient compatibility studies can also be conducted during the preformulation stage to screen for potential incompatibilities.

Formulations can be stabilized by modifications to the composition, process, or packaging. Desiccants can be added to the package to stabilize a moisture-sensitive product. Dry granulation also can be used instead of wet granulation to prevent moisture instability. One simple way to stabilize a formulation is to keep the excipient/active ratio as low as possible. Formulation additives such as antioxidants and chelating agents can also be added for stability. It is also very important to understand the properties of excipients. For example, a slurry of granular dibasic calcium phosphate dihydrate will produce a pH close to neutral. On the other hand, a slurry of granular dibasic calcium phosphate anhydrous has a pH near 5. Thus, the dihydrate form of dibasic calcium phosphate may be a better choice for formulating an active ratio that is unstable at low pH.

Stability studies are conducted with the goal of having degradants controlled to within the International Conference on Harmonization (ICH)-established limits for drug products at the time of the NDA filing (9). These limits vary according to the daily dose, and higher limits are allowed with qualification studies.

Summary

QbD, as the name states, means that every aspect of a drug product has been purposely built into the finished dosage form to yield a product of highest quality. Building a formulation development portfolio from the ground floor through science-driven approaches to solid form selection and formulation development is the first step to producing a commercial product that is stable, bioavailable, and globally acceptable. This approach will form the foundation for a successful FDA PAI.

IN VITRO/IN VIVO CORRELATION IN EARLY DRUG DEVELOPMENT

Understanding the relationship between in vitro properties and in vivo performance of a dosage formulation early in the drug development process is both beneficial and practical. Whether this relationship is (1) a quantitative, validated, predictive model [in vitro/in vivo correlation (IVIVC)], (2) a qualitative, possibly semiquantitative association [in vitro/in vivo relation (IVIVR)], or (3) no correlation at all between in vitro and in vivo data is dependent on several factors:

- The BCS class of the compound may indicate feasibility of establishing an IVIVC or an IVIVR.
- The intended design of drug release from the dosage formulation may indicate what sort of relationship is possible (typically an IVIVC for modified release and an IVIVR for immediate release).
- An IVIVC may still not be acceptable as a surrogate for bioequivalency testing if the drug is considered to have a narrow therapeutic index.
- The degree of linearity in pharmacokinetics and/or variability in rate and extent of absorption may limit the correlation.
- The in vivo release of the drug from the dosage formulation should be relatively insensitive to range of variation expected within the in vivo environment (gastrointestinal pH, agitation, transit, surfactants).

The rationale for developing an IVIVC is to allow in vitro dissolution to serve as a surrogate for in vivo bioequivalence. Prediction of in vivo performance based on in vitro dissolution establishes the dissolution method as biorelevant. The correlation may be used in the justification of dissolution specifications, and the correlation validates the use of dissolution testing and specifications as a quality control tool for process control.

There are several practical considerations pharmaceutical scientists can apply early in drug development to assist in establishing a correlation:

- Develop a series of dosage formulations that are similar in their mechanism of release, with release rate differences due to one, or possibly two, rate-controlling critical manufacturing variables.
- Obtain human in vivo data as early as possible to aid in the direction of formulation development. The sooner a correlation can be established, the sooner the appropriate dissolution method is used to begin assembling historical database and stability information on the dosage formulations.
- Make use of design of experiment (DOE) approaches when investigating in vitro test conditions to obtain more relevant information with reduced time and use of resources.
- Use three dosage units per condition, instead of the traditional six, to save time and resources when initially screening dissolution test conditions.

- Use the differences in the in vivo absorption profiles to direct in vitro test development. Often, it is not the absolute value of the dissolution profile that correlates with the in vivo data, but the relative degree of difference among formulations that make the dissolution test conditions discriminating.
- Treat each case on an individual basis. Success with one drug formulation may not be extrapolated to all drugs or to different formulations of the same drug.

The availability of an IVIVC can impact the drug product development process and time lines in a number of areas, from the initial dosage formulation development, to dosage formulation optimization and in-process controls, and finally to post-approval qualifying changes. The development of correlations has evolved to become a dynamic process that should be considered from the very outset of dosage formulation development, and then continually monitored and adapted throughout the integrated process of formulation development and dissolution method development. Timely meetings with the FDA (e.g., end of phase 2) to review with, or to solicit guidance from, the agency on the IVIVC can prove beneficial in the development process and ultimately aid in the PAI and the NDA review processes.

CONVENTIONAL ORAL DRUG DELIVERY

During the early development of a new chemical entity, the development program is focused on achieving a formulation that is bioavailable. As the program progresses into later stages of clinical development, the focus changes into developing a formulation that is stable and a manufacturing process that is robust. A stable formulation and a robust manufacturing process will lead to a quality product, but getting this formulation to those simple endpoints requires a high level of process knowledge. Various factors to be considered in developing oral dosage forms are described in the pages that follow.

Development of Formulation

At the base of any dosage form is the formulation. The formulation creates the environment where the active ingredient is either in harmony with its surroundings and/or protected from environmental moisture and oxygen. Whenever possible, preventing degrading interactions should be the first priority in creating a stable dosage form.

Once the stable dosage form is created, the formulation needs to be tested with respect to its robustness. The testing is usually done through a statistical design of formulation range study. In this study, all key ingredients are tested to see how they affect the critical product attributes. For example, if the tablet matrix is altered, it may change the potency, uniformity, or dissolution characteristics of the dosage form. Changing the level of certain ingredients, like fillers, may also enhance or reduce the uniformity of the dosage form by making the active concentration either too high or too low to be effectively manufactured on current production compression or filling machines. Likewise, changing glidants or lubricants may change the way the product flows into die cavities that in turn impact uniformity or dissolution characteristics of the final dosage form. Given the number of ingredients in a dosage form, and the degree of change for those ingredients, fractional factorial experimental designs are usually performed to determine the major and minor effects of the ingredients on the tablet matrix.

When running designed experiments on formulations, the scientist must be ready to encounter unexpected interactions due to the different components in the system. An example of discovered formulation interactions might be something as simple as the use of sodium lauryl sulfate (SLS) in the formulation. Surfactants like SLS are often used to enhance dissolution characteristics, but formulation range studies have shown that the presence of SLS in the formulation can decrease dissolution characteristics, or enhance lubricant properties of the formulation. These effects are certainly unexpected; however, they do exist, and if discovered early enough in the formulation development program, give the scientist the ability to enhance or improve the formulation to ease future manufacturing difficulties.

The Manufacturing Process

If the manufacturing process is not optimized, it could decrease the manufacturing design space, and create a product that cannot be scaled up to the production scale. The goal of process development is to create manufacturing processes that are both simple and robust.

Most of the solid dosage manufacturing processes, independent of the number of manufacturing steps, have the same characteristics: the process must create a uniform granulation; that granulation needs to be delivered to the tablet press or capsule filler uniformly; and the tablet press or capsule filler must be able to control the final weight of dosage form to ensure delivery of a bioavailable active to the patient. These critical steps must be controlled whether the process is a simple dry blend, roller compaction, or wet granulation process.

The way to start a process development program is to look at the formulation and ask: What affects the formulation itself? Will it be susceptible to oxygen, moisture, or heat? Additional questions relate to the physical properties of the formulation. Do the ingredients flow by themselves or will they need help? If the formulation flows, will it fill into capsules or compress easily without affecting dissolution? If the formulation is stable, flows well, and compresses easily, then the potential for a direct compression process exists. This type of development keeps the process simple, which will make the final transfer to the manufacturing facility more efficient.

The manufacturing process should be kept as simple as possible. The simplest one is the dry blend process. If the ingredients can be mixed uniformly but need a little help to flow, the addition of a roll compaction step can be added. The roller compactor can change the blend into a stable granulation that can enhance flow characteristics. A roller compaction process, when fed a uniform blend can also lock the active ingredients into new mesh fractions that can help with delivery of the granulation to the tablet press. This type of process addition can make the formulation more robust which will improve the manufacturability of the product while improving quality. The addition of a roller compaction step means that the new parameters, such as compaction, must be monitored and controlled along with dry sizing, but the improvements in process robustness usually outweigh the extra processing time required to perform this manufacturing step. A typical process flow diagram involved in the dry blend process using roller compaction is illustrated in Figure 1.

With the addition of the roller compaction step, the level of testing required to evaluate the performance of the process increases. The design of experiments will also expand to meet the added requirements of the process.

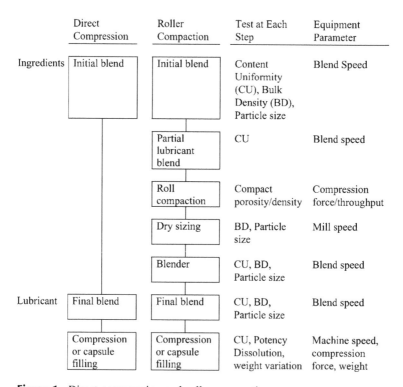

Figure 1 Direct compression and roller compaction processes.

Wet Granulation Process

If formulation problems are too extensive to solve by blending, or roller compaction, the process may need the problem-solving capability of a wet granulation process, which is more adaptable and forgiving. The wet granulation process, like fluid bed granulation or high shear granulation, can take poorly compressible product, or a formulation with low solubility or stability, and change the granulation by linking binders, surfactants, wetting agents, or preservatives directly to the active granule. With the judicious selection of ingredients, the granulation usually achieves an improved level of stability and improved compressibility that can help the formulation as it passes through the tablet press or capsule filler.

The downside of the wet granulation process is that the extent of granulation development usually plays a major role in compression, uniformity, potency, stability, and dissolution characteristics. This is due to a variety of characteristics of wet granulated material such as compressibility, porosity, bulk density, and flowability. All of these critical product attributes need to be monitored, and process optimization will have to be performed to mitigate the new risks. There is always a need to monitor granulation endpoint either by its physical characteristics or its impact on the processor while developing the granulation. The endpoints are easy to monitor with a high level of visual acuity training by the scientist, but the endpoint of development can also be monitored by standard technology solutions such as power endpoint, torque, or acoustics.

When developing a manufacturing process, whether it is by direct compression, roller compaction, or wet granulation, a good understanding of the process conditions is required. A new process range study will always need to be run for each new formulation or process.

Once the formulation and process are defined, it is time to run the process range study. This step is best accomplished by running a second statistical design study. For this study, the formulation (ingredients) will stay constant, and the process will be manipulated to determine the impact of critical unit operations on the final tablet characteristics. Simple processes, like direct compression, may have designs with three to four critical factors for the process.

With the addition of a roller compaction or wet granulation process, the evaluation becomes much more complicated because any of the new manufacturing steps can change the parameters of the final blend, delivery, and compression characteristics of the finished product. Examples of cause and effect relationships are shown in Table 2. For example, the extent of granule development can change particle size, which can itself affect product flow, uniformity, potency, compression, or dissolution independent of the final blend lubricant effects or extragranular ingredients. All of the potential effects have to be evaluated in order to determine their individual significance to the product. Once the process is fully understood, the process scientist has the ability to manipulate the individual steps to achieve the desired critical quality attributes of the finished dosage form.

Table 2 Cause and Effect Relationships

Evaluated step	Can affect	Evaluation parameter
Initial blend step	Dissolution Potency Uniformity	Bulk density Particle size Granular flow
Granulation development step	Dissolution Potency Uniformity Stability	Bulk density Particle size Granule porosity Granular flow
Milling step	Dissolution Potency Uniformity	Particle size Bulk density Granular flow
Final blend step	Dissolution Uniformity	Extent of lubrication Granular flow
Compression step	Potency Uniformity Disintegration Dissolution	Applied compression force Press dwell time
Drying step	Dissolution Stability	Moisture content

If a formulation cannot be manufactured using conventional blending or granulation processes, the scientist can move on to more involved manufacturing processes, i.e., active overcoating process. Pharmaceutical scientists can develop advanced techniques once they understand the relationship of the drug's physicochemical properties to product quality. Active overcoating is a technique which can be used successfully for highly bioavailable active ingredients that have product stability issues. The manufacturing process includes preparation of the active suspension or solution, which is then sprayed onto specially prepared inert-filled placebo tablets. The coated inert-filled placebo tablets are subsequently colored, polished, and finally branded.

CONTROLLED DRUG DELIVERY

On the basis of the FDA's critical path initiative (1) to modernize the drug development process, Wyeth volunteered to initiate development of one of its new drug products as part of this pilot program. The FDA partnered closely with us throughout the development, filing, and approval process. On the basis of the Pharmaceutical cGMP for the 21st Century directive, a risk-based approach was adopted. This focus on the product development process is expected to facilitate risk-based regulatory decision-making, innovation, and use of risk identification and control strategies. Overall, the approach resulted in enhancements to the CMC section of the NDA, including

- Demonstration of product knowledge and process understanding through scientific rigor
- QbD, i.e., build quality in the product
- Information on critical quality attributes and their relationship to clinical performance (safety and efficacy)
- Identification of critical and noncritical parameters and the associated quality attributes, with the development of design and control space for critical factors
- Discussion regarding use of PAT applications as a means to control variability and thus as a risk mitigation strategy. With enhanced understanding of sources of variability in the manufacturing process—groundwork was laid for future use of PAT for real-time release of product
- Identification, mitigation, and control of risks associated in the process
- More reliance on the Quality Overall Summary document (module 2) as a comprehensive summary document that captures the information, knowledge, and understanding of the drug substance/product throughout the development cycle

As an example, a controlled-release dosage form was developed using a drug substance of the BCS class 3. In retrospect, earlier clinical formulations (through phase 2 and into phase 3) using a high shear wet granulation manufacturing process to enhance densification and flowability would have been a more rigorous example. Subsequently, a more efficient, simpler yet robust roller compacted process was selected for commercialization, with no change in formulation composition. The granulation is compressed and film-coated to yield the final dosage form.

Of significance, an IVIVC was demonstrated for the matrix formulation, thus dissolution is a surrogate marker for clinical performance. The dissolution range defined by the IVIVC established the boundaries of the bioequivalent design space. Having an IVIVC early in the program helped in development of different formulations to bridge the process change (wet granulation to roller compaction) to gain advantage in time and regulatory flexibility. The formulation consists of a high molecular weight polymer as a matrix controlled released tablet. The dissolution was based on the polymer concentration and the specific surface area of the tablet.

On the basis of prior knowledge and understanding of the formulation and process, a quality risk assessment was performed to understand the criticality of different formulation and process variables. Variability, with respect to physicochemical and functional properties of all excipients used in the formulation design, was well understood. On the basis of the risk assessment, tablet dissolution and tablet hardness were identified as critical quality attributes. Polymer concentration was a critical factor that impacts dissolution. Polymer properties, on the other hand, played a minor role in the dissolution performance. Control of polymer concentration within the design space yielded consistent dissolution performance. The impact of variability in drug substance particle size and morphology was extensively studied and demonstrated to have minimal impact on dissolution.

From a processability (manufacturing) aspect, tablet hardness was identified as the critical quality attribute. The roller compaction parameters were identified as critical parameters that impact granule characteristics and subsequent compressibility (tablet hardness).

With the QbD approach, appropriate multivariate experimental plans were designed on the basis of this prior knowledge and risk analysis. This DOE approach established the design space for the critical parameters through a series of experiments that were conducted at a development scale. During the DOE, the influence of varying manufacturing parameters on the formulation and process was investigated to verify the process parameters with greatest impact on quality. The design space was developed using pilot scale batches, and the design parameters were developed using known scale-up factors provided by the equipment manufacturer. The predicted design space was subsequently verified using larger scale batches, which included the primary stability lots. Operation of the manufacturing process within the design space is expected to produce drug product of consistent quality. The cumulative effect of all noncritical factors based on unit operation was also reviewed to understand any downstream effect on manufacturing process.

PAT tools were used to understand and control the variability of the manufacturing process at five different places on the basis of the experience gained with respect to critical parameters. PAT tools designed to measure the blend uniformity at both the initial and final stages and to measure tablet content uniformity were developed using a near-infrared spectroscopy technology. The particle size of the roller compacted granules are monitored in real-time by use of a laser diffraction particle size instrument. With sufficient validation and batch history, it is proposed that these PAT tools could facilitate eventual real-time release (i.e., no end product testing). PAT tools measure important characteristics such as tablet weight (shown to be suitable for uniformity of dosage units), active drug, and release-controlling polymer levels (linked directly to potency and dissolution performance).

A constant dialogue was maintained between the FDA and the Wyeth product development team to update and consult on the basis of the pilot program initiative. This helped in gaining concurrence with the agency in our approach toward the submission because of transparent communication (building trust), and demonstrated our enhanced mechanistic understanding of the product and process. This helped us tremendously during the actual PAI, where there was a greater emphasis on science, resulting in very few CMC-related questions, and the focus was primarily on plant quality systems.

NOVEL DRUG DELIVERY SYSTEMS

Wyeth has successfully collaborated with other organizations in developing novel dosage forms and delivery systems. One such example is the drug eluting stent. Stents are metal tubular devices used to treat blockages in arteries. A common failure of metal stents is the occurrence of tissue proliferation at the stent site, which causes blockage and generally requires replacement. Drug-eluting stents are normal metal stents that have been coated with a pharmacologic agent (drug) that is known to interfere with tissue proliferation, preventing, or reducing reoccurrence of blockage. Prior detailed knowledge of physical and chemical properties of the chosen drug was critical in the development of a successful product. This included understanding the following:

- Stability behavior and likely degradation pathways of the different crystalline phases of the drug
- Solubility of the drug in aqueous and organic solvents
- Understanding of stability of the drug in various solvent systems
- Drug release mechanism

The coating process used in the manufacture of this drug-eluting stent resulted in the drug being deposited in its amorphous state. This was advantageous for the device, since amorphous drugs have properties that enhance performance. In general, amorphous drugs have a higher solubility, which in this case will increase the drug release rate of the stent. The amorphous drug will also form a more consistent, uniform coating on the stent. However, there are several drawbacks to using amorphous drugs, which fall into two categories, physical stability and chemical stability. Physical stability is a concern because it is not unusual for amorphous drugs to crystallize over time. Crystallization could reduce the release rate of the drug from the stent and change the physical appearance of the stent. The chemical stability of amorphous material is typically reduced when compared with the crystalline drug. An amorphous drug is also more likely to absorb moisture. These issues were overcome by detailed knowledge of the chemical and physical properties of the drug and device and the development of product-specific analytical methodologies. In this case, success was achieved by combining both the pharmaceutical and the medical device scientist's knowledge.

PARENTERAL DRUG DEVELOPMENT CHALLENGES

The development of drug products and processes for parenteral drugs poses a number of unique challenges for the pharmaceutical scientist. Since many parenteral drugs must spend at least some time in a liquid environment prior to entry into the body, poor solubility and stability of a drug in solution typically requires special formulation approaches. Aqueous solubility is an important factor for all drugs, but can be particularly important when designing drugs for intravenous administration. New drug candidates entering development are increasingly larger and lipophilic, and both factors reduce aqueous solubility (10). Administration of drugs by injection into the body is inherently more risky compared with other routes of administration because the natural defense barriers of the body, e.g., skin and mucous membranes, are circumvented during administration. As a result, the toxicities of excipients and other added substances may have a more limiting effect on the acceptability of a particular formulation. Therefore, the parenteral product development scientist has a more limited choice of acceptable ingredients to work with in order to overcome difficulties in drug solubility and stability.

All parenteral products must possess characteristics of sterility, freedom from pyrogens, and extraneous particulate matter (11). Therefore, once the pharmaceutical scientist selects a formulation, it must be processed and handled in a manner that permits these characteristics to be imparted to the final product. Most commonly, the method of sterilization that is selected for a dosage form is dependent on the physical and chemical stability of the drug and other components of the dosage form. For example, terminal sterilization provides the greatest degree of sterility assurance for parenteral products, and is the most desired way to process sterile products. However, the physical and chemical lability of the drug and physical characteristics of the formulation may preclude the use of heat or radiation as methods of sterilization. In addition, contact materials that are encountered during storage, processing, and handling of parenteral products are a principal area of concern, and much study and evaluation of contact materials typically occurs during product development. Thus, the parenteral product pharmaceutical scientist must possess an understanding of material characteristics, including physical and microbiological, as well as container closure integrity.

The advantages of parenteral administration versus other routes of administration are well known (12). From a drug development standpoint, one advantage is the flexibility of dosing in clinical trials, which may help simplify and reduce the number or amount of clinical supplies. By careful planning and continuous study of drug and dosage form characteristics throughout the period of clinical development, a dosage form that is developed for early evaluation in animals and man is more likely to be successfully developed into a full scale commercial product versus an oral dosage form. Several examples of challenging parenteral development projects will be discussed below.

Liquid Parenteral Product Development

The development of a liquid parenteral product is discussed in the context of QbD. For example, an active ingredient with low water solubility and the potential to degrade via hydrolytic and oxidative pathways can be quite challenging to formulate. In one such case, preclinical and clinical studies required that a wide range of potential doses spanning more than two orders of magnitude be accommodated by the formulation.

General formulation approaches that were considered to produce a parenteral dosage form included sterile solid, via lyophilization or powder fill, or a nonaqueous liquid concentrate employing a combination of cosolvents/ surfactants and stabilizing agents. The latter approach was selected because of the simplicity of processing and proved successful in producing a dosage form that provided the required flexibility in dosing. In addition, the liquid formulation could be produced using conventional manufacturing equipment at both clinical and commercial manufacturing sites. This approach facilitated development through small- and large-scale manufacturing trials. Experience gained through the manufacture of small-scale batches proved highly valuable when transfer to the commercial manufacturing site occurred because of general similarities in process at each scale.

Because the active ingredient is potentially susceptible to various routes of instability, a comprehensive understanding of the physical and chemical properties of a drug was important to help evaluate how formulation, process, and environmental factors affect the quality of the drug product. The stability of the drug is affected by numerous factors including acids and bases, metals, and any materials capable of initiating or catalyzing oxidation. Although production of the drug as a liquid provided some process advantages, the role of contact materials and excipient characteristics on dosage form purity and stability needed to be well characterized and controlled because of the potential presence of catalysts such as metals and organic extractables. In addition, environmental factors, such as moisture levels, oxygen levels, and light exposure, were also studied.

A QbD approach was applied to the formulation and process development of this parenteral product. The critical process parameters were identified and separated from noncritical parameters or factors that have minimal influence on product quality. The parameters that were included in QbD studies included process and environmental factors, as well as raw material characteristics. A list of potential risk factors was developed, and the likelihood each factor would contribute to a loss of product quality could be evaluated and ranked. Factors that have a low risk of adversely affecting product quality did not need strict control or remediation, while more significant process controls could be adapted for the more significant risk factors. Experiments were also designed to examine potential interactions among the most significant factors. Establishment of a design space for those factors that significantly affect drug product quality helps ensure a consistent product.

Enhancement of Product Attributes Critical for Parenteral Product Quality and Clinical Administration

Another example of the development of a lyophilized parenteral product is discussed in the context of product attributes needed for clinical administration. The initial formulation met the subvisible particulate matter criteria as specified.

However, routine product performance surveillance indicated sporadic occurrences of excessive particulate matter. As an interim measure, adjustments to the manufacturing process and the shortening of the expiration dating provided control over the excessive particulates. In 1995, USP <788> (13) standards for particulate matter were significantly tightened from the previous standards for the 10- and 25-μm particles. This strictness increased the urgency for conducting a root cause analysis for subvisible particulate matter formation in the presence of common commercial diluents used for intravenous admixture preparation and potential factors for chemically induced particulate formation. The findings from the root cause analysis work, using a QbD approach, led to the modification of the product.

Root Cause Analysis

Particulate Matter in Injectable Products

Particulate matter in parenteral products can arise either exogenously or endogenously. Exogenous particulates are those that result from sources other than the drug product itself. Examples of sources include insoluble compounds in intravenous bags and bits of septa and corings that can be introduced when drug containers are pierced with hypodermic needles.

Endogenous particulates arise from the product itself and are generally the result of chemical reactions that occur during processing and storage of the drug product, or preparation of the intravenous administration solutions for patient use. Minimization of endogenous particulates requires extensive testing and analysis of the drug product under all conditions of potential use and adjusting the formulation design or the product manufacturing process accordingly.

Causes of Endogenous Particulate Formation

Endogenous particulate formation in the product solutions for clinical infusion can occur in at least three ways. First the soluble form of the drug salt can convert into an insoluble free acid or freebase when the product admixture solution is prepared using an acidic diluent from a commercial manufacturer.

Second, the combination of the product and acidic diluents can cause low pH shifts. These shifts can affect particulate formation by means of hydrolysis of labile bonds of the compound, which can then react with intact parent molecules to form an insoluble dimeric form.

Third, metals could catalyze the hydrolysis of the same labile bonds to increase the rate of particulate formation. Since the metal ions act only as catalysts for the reaction and are not consumed during the hydrolysis, the presence of a relatively small amount can increase the amount of particulate formation.

Metal ions can originate from a variety of sources, including the product manufacturing equipment, as trace impurities in both the active substance and excipients, and from the devices used to administer the drug in the clinic.

The most significant exogenous contribution of metal ions appears to be from commercial intravenous solutions used to reconstitute the drug product and prepare diluted admixture, as those commercial solutions tend to have widely varying levels. In summary, the highest levels of particulate matter occur when the diluent has either a high level of metal ions or a low pH level. The combination of both may act synergistically.

With this empirical information, the pharmaceutical scientist is better prepared to develop a product that would be robust enough to satisfy the tightened USP <788> particulate specifications under all expected conditions of use.

THE ROLE OF PACKAGING SCIENCE IN PRODUCT DEVELOPMENT

All drug products that reach the market do so with some sort of package, which is chosen to meet the needs of the patient as well as the manufacturer and distributor. The ideal package should

- provide sufficient protection against environmental factors (humidity, light, etc.) to ensure stability and optimal shelf life,
- protect the product from physical damage during shipping,
- be consumer-friendly (size, shape, convenience),
- help ensure patient compliance to the prescribed dosing regimen,
- provide the necessary level of child-proofing, and
- be inexpensive and simple to manufacture.

All of these factors are important in making a decision on the final commercial product package; however, development of "packaging knowledge" is usually not a specific goal during the product development program. A scientific rationale for package choice based on product requirements is an important part of the submission data package and also for manufacturing site PAI readiness. Understanding the role and importance of the package is a critical function of the pharmaceutical scientist: the knowledge gained during development helps ensure that a high-quality product is approved and supplied to the market successfully for many years, and even assists in future product life-cycle optimization.

Protecting the Product

From a formulation perspective, the most obvious reason for drug product packaging is to protect it from the effects of environmental variables such as humidity, light, and oxygen. These variables can lead to changes in the physical and chemical properties and lead to product failure during shelf life. A well-designed stability program will ensure that the right knowledge is generated about the product (14). However, development timelines and resource constraints may result in a commercial product that is over packaged, adding to the product cost-of-goods and perhaps reducing supply chain flexibility.

A primary goal of phase 2/phase 3 product development should be to determine the minimum protection for the proposed commercial product by understanding how the product's critical quality attributes (CQAs) are affected by environmental variables. The CQAs will differ depending on dosage form. For example, in the most common case of a solid oral dosage form (tablet or capsule), the CQAs may be the following:

- *Chemical Stability*: In order to maintain degradation product levels below the ICH or qualified thresholds throughout the desired shelf life, the packaging material needs to be sufficiently protective. Forced degradation and excipient compatibility studies on the API will provide an understanding of the routes of degradation (e.g., oxidative, hydrolytic, photolytic) as well as a basic understanding of the mechanisms and kinetics. Likewise, stability of early clinical formulations provides important knowledge of the potential for degradation. The role of multiple variables and formulation parameters can ultimately help determine the packaging needed for acceptable stability. In addition, for products with multiple dosage units per package such as tablets in bottle, the moisture vapor transmission rate per unit dose need to be considered to provide sufficient protection (15). Cost and machineability may limit the choices of commercially viable packages.
- *Dissolution*: Changes in the physical properties of the active ingredient or the excipients can affect a product's in vitro dissolution (and potentially its in vivo performance) and should be controlled through the choice of packaging. These can include the effect of moisture on disintegrant performance or as a trigger for crystallinity (polymorph) or hydration state changes of the API. A good example of the importance of packaging is illustrated for a product, which was developed using a polymorph, which had lower stability and, therefore, higher solubility. In this situation, it became necessary to demonstrate that interconversion of the polymorph in the finished product was prevented through a choice of the appropriate packaging. When presented with a clear rationale showing how polymorph conversion was controlled in the finished product, the FDA agreed that bioequivalence studies with low levels of the polymorph were not required. Because the impact of physical changes on bioavailability is potentially large, the discriminatory power of the dissolution method toward these changes should be verified during development.
- *Appearance*: Although the physical appearance of a drug product is not necessarily an indication of product performance (efficacy or safety), it is a quality attribute that is important because it implies quality of these items. Clearly, the aesthetic attributes of a product may influence a consumer's or patient's willingness to use the product. The correct choice of a packaging material or system can help prevent undesirable changes in appearance. The most common preventable cause of appearance changes for a solid

oral dosage form is surface discoloration due to light exposure (fading, mottling). Opaque bottles or blister films or use of an opaque secondary package will prevent this.

APPLICATION OF PAT IN EARLY DEVELOPMENT

The application of PAT in early development should follow the same guidelines that would be used to determine PAT opportunities during commercialization. An early risk assessment should be performed (3) on the proposed process, potential risks identified, and mitigation strategies developed. Some of the mitigation strategies will include the use of PAT. Of course, the early risk assessment is based more on potential issues rather than empirical data since there is little or no empirical data available.

The benefits of using PAT early in the development stage are enormous. A full understanding of the process can be developed at the pilot scale. A large database can be generated that can help define processing ranges. This process understanding and knowledge will ease the technology transfer process during commercialization of a product.

In one example, a coating process transfer from one site to another was successful in the first batch (evaluated by dissolution profile) because of the PAT knowledge database developed at the sending site. This knowledge allowed the receiving site to define the coating endpoint, even though different scale equipment and slightly different processing parameters were utilized at the receiving site. Without this prior knowledge, this process endpoint would have needed to be redefined in the new equipment. This would have required several additional evaluation/demonstration batches followed by the traditional three (3) batch validation.

The benefits of the PAT approach are as follows:

- Saving time to market
- Saving money on development and technology transfer by minimizing the number of batches needed to define and transfer your process
- Increase in process understanding
- Reduction of investigations
- Overall better product quality

CONCLUSION

In this chapter, the importance of understanding mechanistic principles of science and technology in the dosage form development is illustrated through various examples, which have proven to be acceptable from a commercial point of view. The awareness of regulatory knowledge is important in managing a successful PAI, which is the last step in the development phase. It is emphasized that the PAI is not an end of the process, but a process to be built right from the beginning phase of dosage form development.

One of the key pieces of documentation, which covers all the formulation development activities from pre-formulation to the final commercial dosage form, is the "Pharmaceutical Development Report." This report is a required element of the quality-module 3 and can play a pivotal role in PAIs. Both the dossier application reviewer and the district inspectorate closely review the document, and a well-written report can avoid many questions and facilitate smoother PAI (16).

ACKNOWLEDGMENT

The authors sincerely wish to thank all the scientists at Wyeth Chemical and Pharmaceutical Development Division for their valuable contribution and special thanks to Mike Yelvigi for his compilation of this article.

REFERENCES

1. FDA. Pharmaceutical cGMPs for the 21st Century: A Risk-Based Approach: Final Report Fall 2004. 2004. Available at: http://www.fda.gov/cder/gmp/gmp2004/GMP_final report2004.htm.
2. FDA. Guidance for Industry: PAT—A Framework for Innovative Pharmaceutical Development, Manufacturing, and Quality Assurance, CDER/FDA, Sept 2004. Available at: http://www.fda.gov/cder/guidance/6419fnl.pdf.
3. ICH. Guidance for Industry: Q8 Pharmaceutical Development. May 2006.
4. FDA. Guidance for Industry: ANDAs—Pharmaceutical Solid Polymorphism. Draft. 2000. Available at: http://www.fda.gov/cder/guidance/6154dft.pdf.
5. Byrn S, Pfeiffer R, Ganey M, et al. Pharmaceutical solids: a strategic approach to regulatory considerations. Pharm Res 1995; 12(7):945–954.
6. FDA. Guidance for Industry: Waiver of *In Vivo* Bioavailability and Bioequivalence Studies for Immediate-Release Solid Oral dosage Forms Based on a Biopharmaceutics Classification System. 2000. Available at: http://www.fda.gov/cder/guidance/index.htm.
7. Inactive Ingredients Database. Available at: http://www.fda.gov/cder/
8. FDA. Guidance for Industry: Nonclinical Studies for the Safety Evaluation of Pharmaceutical Excipients. 2005. Available at: http://www.fda.gov/cder/guidance/5544fnl.htm.
9. FDA. Guidance for Industry: Q3B(R) Impurities in New Drug Products. 2006. Available at: http://www.fda.gov/cder/guidance/7385fnl.htm.
10. Lipinski CA, Lombardo F, Dominy BW, et al. Experimental and computational approaches to estimate solubility and permeability in drug discovery and development settings. Adv Drug Deliv Rev 1997; 23(1):3–25.
11. Akers MJ. Parenteral preparations. In: David T, USIP staff, eds. Remington: The Science and Practice of Pharmacy. Chapter 41. 21st ed. Philadelphia, PA: Lippincott Williams & Wilkins, 2005.

12. Turco S, King RE. Sterile Dosage Forms. 3rd ed. Philadelphia, PA: Lead & Febiger, 1987.
13. The United States Pharmacopeia. USP <788> Particulate Matter in Injections.
14. Carstensen JT, Rhodes CT, eds. Drug Stability: Principles and Practices. 2nd ed. New York, NY: Marcel Dekker, 1995.
15. PQRI Container Closure Working Group. Basis for Using Moisture Vapor Transmission Rate Per Unit Product in the Evaluation of Moisture-Barrier Equivalence of Primary Packages for Solid Oral Dosage Forms. September 2004.
16. Shuren J. Submission of Chemistry, Manufacturing, and Controls Information in a New Drug Application Under the New Pharmaceutical Quality Assessment System; Notice of Pilot Program. Federal Register, 2005; 70(134):40719.

4

Training Requirements in Product Development: A Key to a Successful Pre-Approval Inspection

Tammy Chaney Cullen and Marie Crabb-Donat
Eli Lilly and Company, Indianapolis, Indiana, U.S.A.

FOUNDATION FOR TRAINING REQUIREMENTS

Pharmaceutical companies have an incredible responsibility for and commitment to the well being of patients. As a wedding band is recognized as a continuous symbol of the commitment and interdependency of two people, so is the relationship between a pharmaceutical company and its patients. The pharmaceutical company is linked to its patients because of its ongoing promise to improve the quality of patients' lives. This promise is brought to fruition by individuals within the company who must perform critical tasks while meeting regulatory expectations.

The drug development process is governed by the Code of Federal Regulations (CFR) that outlines good laboratory practices (GLPs) (1), good clinical practices (GCPs) (2), and good manufacturing practices (GMPs) (3); these are generically referred to as GxPs. These requirements were developed to ensure that drug products are safe for human consumption, beginning in the early phase development and continuing through the commercialization and product launch (4). Drug development, unlike manufacturing, is dictated by multiple GxPs. This control provides a more complex environment for meeting regulatory

Table 1 Outline of the Associated GxPs and Their Purpose

Phase	GxP	Purpose of GxP
Early phase development	GLPs	Govern the quality and integrity of the preclinical safety data in pre-IND activities
Clinical trials	GCPs	Ensure the credibility of data, as well as assuring the rights, well being, and confidentiality of trial subjects
	GMPs	In development, ensure SISPQ of the clinical trial product, as well as reproducibility for further testing and commercial production
Manufacturing	GMPs	In manufacturing, ensure SISPQ for commercial production.

Abbreviations: IND, investigational new drug; SISPQ, safety, integrity, strength, purity, and quality.

expectations throughout the drug development process. Table 1 outlines the associated GxPs and their purpose.

Just as there must be a robust strategy for developing innovative new drug products, there is a similar need for an effective and efficient training strategy. Section 21 CFR 58.29 of the GLPs (1) and Section 21 CFR 211.25 of the GMPs (3) require that each individual "shall have education, training, and experience, or combination thereof to enable that (individual) to perform the assigned functions." An effective training strategy goes beyond the regulatory requirements by ensuring that each phase of the development process is performed, documented, and verified as required for a successful pre-approval inspection (PAI).

This chapter will highlight how combining compliance requirements and performance needs provide not only an effective training strategy for those engaged in drug development, but also a robust process for ensuring a successful PAI. Specifically, it outlines

- the need for an effective training strategy with associated training requirements,
- the need for a performance-based training approach within this training strategy, and
- regulatory requirements for training during the development process.

NEED FOR AN OVERALL TRAINING STRATEGY

It is vital that companies develop, implement, and continually follow an overall training strategy. By containing specific training requirements, this strategy provides a map for the product development organization to verify that employees are properly trained and are ultimately working toward consistent goals.

The purpose of the overall training strategy is to

- ensure that all employees have a process to *demonstrate* ongoing proficiency in their job duties and
- provide a structured mechanism to *document* that all employees are qualified to perform their job duties

Demonstration and documentation are two crucial components of a successful training strategy. Demonstration speaks to the need for performance-based training in which employees must successfully complete the required activities for a specific job duty. Documentation of this successful completion is vital for the PAI as well as any subsequent inspection. Going further, it is imperative to ensure that good record retention practices are followed so that this documentation is archived appropriately and easily retrievable during the inspection process.

TRAINING REQUIREMENTS WITHIN THE STRATEGY

The training strategy must clearly outline the necessary training requirements. This level of detail is vital for training program implementation, business support and commitment, and PAI success. The requirements of a successful training strategy include the following, and will be described in more detail:

- A brief introduction of the strategy, which includes a definition of qualification
- The alignment of training with specific job duties, through duty-aligned curricula
- Training development and performance assessment processes
- Expectations for acceptable performance and continued qualification
- Use of a remediation plan if acceptable performance is not demonstrated
- Plan for insufficient performance and disqualification

Qualification

A successful training strategy goes beyond merely training individuals; it ensures that these individuals are *qualified*. Qualification is a term used to describe a process of verifying that employees within the organization—who contribute at varying points of the product development life cycle—can perform critical job duties safely, accurately, and independently. Qualification requires the appropriate level and combination of education, experience, and duty-specific training:

Education + Experience + Duty-Specific Training = Qualification

A successful training strategy should include each of these components. It is critical that only those employees engaged in new drug development who possess the appropriate level of education and experience are assigned to perform these duties. But remember, qualification goes beyond education and experience; it also includes duty-specific training.

To identify duty-specific training needs, begin by analyzing a job. A job is a specific role, which encompasses all of the significant areas of work assigned to the individual performing the job. A duty is each significant unit of work an individual may be assigned as part of this job. The training strategy includes identifying necessary job duties and the knowledge and skills, or competencies, needed to perform these duties.

Duty-Aligned Curricula

The training strategy must align curricula to these duties accordingly, through the use of duty-aligned curricula assignments. Aligning duty-based curricula is the process that involves

- analysis of the overall job to identify all duties,
- identification of associated tasks required by each duty,
- identification of existing training associated with these tasks,
- gap analysis, as needed, for tasks without associated training, and
- creation of manageable "chunks," or "bundles," of learning associated with each duty.

These manageable chunks, or bundles, of learning, then, become the specific duty-based curricula that an employee can complete to be deemed qualified to perform that particular duty independently.

Training Development and Performance Assessment

The next requirement of a successful training strategy focuses on training development and the associated performance assessment processes. As part of the process of identifying job duties and aligning curricula accordingly, the knowledge and skills, or competencies, needed to perform that duty are identified. Those competencies are reviewed and approved by senior employees with experience in the area [subject matter experts (SMEs)]. Measurable objectives (5) are then developed, that describe the exact competency (performance) employees must exhibit, under what conditions they must exhibit this performance, and the acceptable level of performance.

Assessment (testing) materials must match the objectives. Employees must complete the performance assessment to the degree stated within the objectives. Often the same SMEs serve as qualified instructors, who teach and evaluate employees' performances. The flowchart in Figure 1 illustrates the employees' progression through the assessment process.

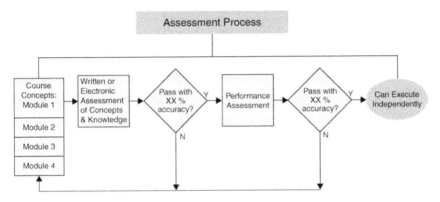

Figure 1 Performance assessment process.

In summary, we identify the competencies product development employees must have, provide performance-based training solutions for those competencies, and evaluate employees' performance in order to obtain duty-specific qualification.

Acceptable Performance and Continued Qualification

The training strategy should contain requirements regarding acceptable performance and continued qualification. To become a qualified employee for a specific duty, *all* acceptable performance standards must be met for *each* of the critical job duties. When this occurs, an employee will have evidence that he or she has completed the assigned duty-aligned curricula.

In order to maintain qualification status, the employee is required to continually perform his or her job duties correctly, safely, and independently. The employee's ongoing performance of a duty will be monitored through the following means:

- Observation of employee performance
- Qualitative assessment of feedback acquired from colleagues, support staff, and supervision regarding any employee issue
- Ongoing compliance to and completion of duty-aligned curricula

Remediation Activities

The next requirement within the training strategy outlines the plan for remediation. If employees are not able to meet acceptable performance criteria for a given job duty in the development of new drugs, they will work with the qualified instructor in a one-on-one remediation activity. During remediation, the employee will receive additional instruction, practice time, and coaching from the qualified instructor (Fig. 2).

After remediation occurs, and when both the employee and qualified instructor are confident that the tasks can be performed correctly, safely, and independently, the employee demonstrates for performance assessment again.

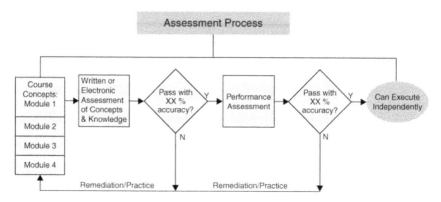

Figure 2   Remediation process.

Employees must complete the performance assessment to the criteria stated within the objectives and to the satisfaction of the qualified instructor. If an employee still does not meet the acceptable performance criteria, the person may be able to execute that task or duty, but only under direct supervision of a qualified instructor. The employee cannot execute that task independently until a successful performance demonstration has been documented with the qualified instructor.

Plan for Disqualification and Re-qualification

Finally, a successful training strategy must include a plan to address insufficient performance that results in disqualification. Disqualification occurs when a person who was once qualified is no longer able to perform acceptably. This disqualification can occur because of many reasons, including

- recurring error,
- supervised observance of performance gaps,
- employee admittance of performance gaps, and
- leave of absence that results in time away from a particular duty.

Re-qualification is the opportunity to become qualified again. It can address significant changes to existing processes or equipment, or individual performance gaps. Depending on the need, the re-qualification may be required for the entire job function or for a specific duty. The employee must complete the necessary training or remediation activities and execute tasks according to the acceptable performance criteria to become re-qualified.

In summary, each of the above noted requirements are crucial to a robust training strategy, and ensures that employees developing and making medicine are, in fact, qualified to do so. Remember, it is this effective and efficient strategy, which is developed, implemented, and adhered to, that aids in a successful pre-approval inspection as well as supports the ongoing commitment to our patients.

PERFORMANCE-BASED TRAINING APPROACH

If you recall, the training strategy has two purposes: demonstration and documentation. Demonstration is defined as an ongoing proficiency in job duties—achieved through the use of a performance-based training approach. The industry is moving away from the traditional compliance-driven training approach toward a focus on performance-based training. Therefore, it is important to distinguish between compliance-driven training and performance-based training (Table 2) (6,7).

To summarize the key principles, a "check the box" mentality is often associated with a compliance-driven training approach—to provide a training solution (whatever that might be), have employees complete the training, and be sure they document their completion. Following are major problems with this type of training program that can be seen from the recurring 483 observations from 2004 to 2007 on the FDA Web site (8):

- Lack of detailed training strategy to adequately train employees on required job activities
- Inadequate training for job-specific activities and associated regulatory requirements
- Lack of sufficient training in GMP regulations
- No provision for ensuring that employees were trained before performing job functions
- Deficiencies in training documentation; including incorrectly scored tests and incomplete or missing signatures

A performance-based training approach ensures that the solution achieves the desired end result (i.e., change in knowledge, skill, or competency) (9). The goal is to develop a plan that ensures individuals can perform the assigned job by translating required documents [i.e., standard operating procedures (SOPs) and work instructions, etc.] to their daily duties. Performance-based training goes beyond compliance by bundling duty and regulatory requirements, with a focus on executing the duty in a manner that is appropriate and safe, within a predetermined degree of consistency and integrity. In other words, this approach focuses on performance that can directly impact the safety, integrity, strength, purity, and quality (SISPQ) of the product and safety of the patient. Employees can then comprehend what they need to do and why the given parameters are important.

Therefore, when creating the training strategy's subsequent training programs, be sure to

- identify the end goal,
- break the goal into smaller, more manageable chunks of learning,
- identify what individuals must know and do to perform job duties,
- bundle performance needs into training programs that allow employees the opportunity to demonstrate performance, and
- ensure alignment between course objectives, assessment activities, and content.

Table 2 Comparisons and Contrasts of the Key Principles Within Compliance-Driven Training Approach and Performance-Based Training Approach

Key principle	Compliance-driven training approach	Performance-based training approach
Driver for change	Change is externally driven through • external regulations and regulatory agencies • legal liability	Change is internally driven through • analysis of present and desired level of performance • identification of causes for performance gaps
Primary focus	Prove compliance by documenting, tracking, and proving the training solution took place.	Prove performance by implementing an appropriate solution in order to bring about a change in knowledge, skill, or competency.
Indicator of success	Success is date driven. The training solution is viewed as successful if it is implemented by the intended date.	Success is performance driven. The results of the solution are evaluated to ensure the intended performance gap is closed.
Range of solutions	The compliance requirement or commitment is assessed, resulting in inflexible interventions that are viewed as obligatory versus value added.	The need is assessed, offering a wide range of interventions with which to improve performance, taking into account work capacity and environmental limitations. The solution is viewed as value added and guides the change management process.
Complexity	While this approach is less complex to develop and implement, the business processes become complex because performance issues are handled in isolation.	While this approach is more complex to develop and implement because of analysis and a wide range of interventions explored, the business processes become less complex because a systems view is taken to focus on outcomes.

Source: From Refs. 6, 7.

The imperative point to remember when developing, implementing, or evaluating the training strategy is that it establishes a performance-based training approach while *maintaining* compliance. It is not a battle of one versus the other. Performance-based training solutions must still be documented, because the second purpose of the training strategy was documenting that all employees are qualified to perform their job duties.

A performance-based training approach, with the appropriate level of documentation, provides verified results for a PAI. If asked during the inspection how employees are qualified to execute the duties within their jobs, you can easily refer to the training strategy in place and the completion of the various components of their qualification—documentation of their education, experience, and completion of specific duty-based training through demonstration of desired performance.

TRAINING REQUIREMENTS WITHIN PRODUCT DEVELOPMENT PHASES

Using the pharmaceutical product development matrix below, it is important to identify the application of the training strategy to the phases within the product development life cycle, including appropriate types of training and associated documentation (Fig. 3).

Throughout the drug development process, a training strategy should be in place to address the changing regulatory requirements. Regardless of the phase for drug development, it is important to ensure that employees are qualified throughout this entire process. However, the stringency of the duty-specific training may vary. On the left side of this process, in Discovery, there is less duty-specific training, as individuals are governed more by "good scientific judgment," than by GxPs. Here, their qualification may be more heavily weighted on education and experience. However, as the drug development requirements become more and more regulated by GxPs, duty-specific training becomes more stringent and robust.

The regulations indicate that individuals must "have the appropriate education, training, and or experience to perform the assigned function" (3).

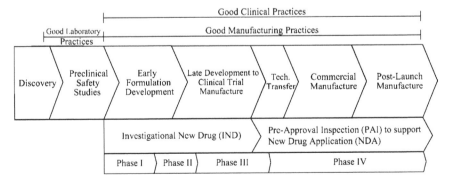

Figure 3 Pharmaceutical Product Development Matrix. *Abbreviations*: IND, investigational new drug; PAI, pre-approval inspection; NDA, new drug application. *Source*: From Ref. 10.

A training strategy that ensures the "ability" to perform focuses on job duties as well as the associated processes required (i.e., SOPs and work instructions) to execute a duty or task.

During early phase development, GLP obligations require "a current summary of training and experience and job description for each individual engaged in or supervising the conduct of nonclinical laboratory study," Section 21 CFR 58.29 (1). In addition, "personnel shall take necessary sanitation and health precautions designed to avoid contamination of test and control articles and test systems" (1). The GLPs provide standards for the planning, performance, monitoring, recording, and reporting of preclinical safety studies conducted in support of an application to market a new drug (11). Training requirements, therefore, include an overview of the process, safety considerations throughout the process, and documentation requirements for each of the process steps.

During clinical trials, GCPs provide the "ethical and scientific quality standard for designing, conducting, recording, and reporting trials that involve the participation of human subjects" (2). Training requirements include an overview of the clinical trial process, patient safety considerations and requirements, and documentation requirements.

GMP guidelines must also be met during all phases of clinical trials. GMP requirements, outlined in Section 21 CFR 211.25 (3), add the requirement of training on "particular operations that the employee performs and in current good manufacturing practice (including ... written procedures ...) as they related to the employee's functions." Training requirements include relevant procedures, work instructions, and duty-specific training.

Training requirements must include the following, as they relate to specific duties:

- Regulatory training
  - Regulatory requirements in GLPs, GCPs, and GMPs and how to incorporate into specific duties
  - Quality principles, including batch disposition, stability, process validation, deviations, management notification, change control, training requirements, and documentation
  - Aseptic operations
  - Rights and safety of patients
- Safety requirements
  - Occupational safety and health administration
- Duty-specific requirements, for example
  - Validation/qualification protocols for manufacturing process, equipment, cleaning, etc.
  - Development history
  - Analytical method writing and validation

- How to set up protocols
- How to report data
- Toxicology, medical requirements
- Company specific equipment, i.e., lab equipment and lab processes
- Understand issues with cross-contamination and cleaning
- Local SOPs or guidance documents that are to be followed for consistency
- Inspection readiness
  - Overview of the PAI strategy
  - Process for providing needed documentation for PAI submission [investigational new drug and new drug application]
  - How to prepare for PAI
  - How to manage a PAI inspection, including how to interact with inspectors

How then, does an organization remain competitive and continue to succeed in an ever-changing landscape?

SUMMARY

The pharmaceutical industry exists in an environment of continued flux, including changing regulatory requirements, economic conditions, and patient needs. Two factors contribute to the success. First, a robust drug development strategy for creating innovative new drug products is critical. Second, and equally important, is an effective and efficient training strategy that supports the drug development process. Together, these strategies create the foundation for meeting our commitment to our patients and ensuring a successful PAI.

ACKNOWLEDGMENTS

A special thank-you to Melynda Buher and Jeanette Buckwalter for their editorial assistance. Your help was greatly appreciated.

REFERENCES

1. The Code of Federal Regulations. 21 CFR Part 58. Good Laboratory Practices for Nonclinical Laboratory Studies. http://www.accessdata.fda.gov/scripts/cdrh/cfdocs/cfcfr/CFRSearch.cfm?CFRPart=58
2. Guidance for Industry E6 Good Clinical Practice: Consolidated Guidance. http://www.fda.gov/cder/guidance/959fnl.pdf
3. The Code of Federal Regulations. 21 CFR Part 211. Good Manufacturing Practice for Finished Pharmaceuticals. http://www.accessdata.fda.gov/scripts/cdrh/cfdocs/cfcfr/CFRSearch.cfm?CFRPart=211

4. Hynes MD, Buckwalter J. The evolution of the Food and Drug Administration: pre-new drug application approval inspection. In: Hynes M, ed. Pharmaceutical Pre-Approval Inspections: A Guide to Regulatory Success. 2nd ed. New York, NY: Marcel Dekker, 2007.
5. Mager R. Preparing Instructional Objectives: A Critical Tool in the Development of Effective Instruction. 3rd ed. Atlanta, GA: The Center of Effective Performance. 1997.
6. Srinivas H. Foster a Learning Culture: Handbook of Instructional Technology. New York, NY: McGraw-Hill, 1993.
7. Webber AM. Will companies ever learn? Fast Company. 2000:39.
8. FDA Web site. Available at: http://www.fda.gov/foi/warning.htm.
9. Cannon MD, Witherspoon R. Actionable feedback: unlocking the power of learning and performance improvement. Acad Manage Exec 2005; 19(2):120–134.
10. Ray L. Equipment cleaning during pharmaceutical product development and its importance to pre-approval inspection. In: Hynes M, ed. Pharmaceutical Pre-Approval Inspections: A Guide to Regulatory Success. 2nd ed. New York, NY: Marcel Dekker, 2007.
11. Hynes MD. Compliance requirements during the drug development process. In: Medina C, ed. Drugs and the Pharmaceutical Sciences: Compliance Handbook for Pharmaceuticals, Medical Devices, and Biologics. New York, NY: Marcel Dekker, 2003.

ADDITIONAL RESOURCES

Mager R, Pipe P. Analyzing Performance Problems: Or, You Really Oughta Wanna: How to Figure Out Why People Aren't Doing What They Should Be, and What to Do About It. Atlanta, GA: The Center of Effective Performance, 1997.
Repenning N, Sterman J. Nobody ever gets credit for fixing problems that never happened: creating and sustaining process improvement. Calif Manage Rev 2001; 43(4): 64–68.
Svenson R. Winning every time: six ways to make large-scale performance interventions succeed. Perform Improv 2004; 23(3):28–32.

5

The Systems-Based Pre-Approval Inspection

Elizabeth M. Troll

*Otsuka Pharmaceutical Development & Commercialization, Inc.,
Rockville, Maryland, U.S.A.*

INTRODUCTION

In the past, the goal of the pre-approval inspection had been to verify that the data supplied by a sponsor in the marketing application were supported by facts. Sponsors would spend many hours organizing and cataloging data and executing data retrieval drills. A majority of these activities were paper based. There was an expectation, on the basis of the volume of information generally contained within an application, that some, if not most, information would be not reviewed during a site inspection. Pre-approval inspection strategies were based on assuring that the most important information (studies that supported the primary claims of the application) was accurate and available during an inspection.

Two factors have rendered this practice obsolete: the advent of the electronic common technical document (eCTD) format for marketing applications and the systems based inspection.

THE eCTD AND ITS EFFECT ON THE SEARCHABILITY OF DATA

The introduction of the common technical document (CTD) was designed to aid sponsors in presenting the data that supported their marketing applications in an organized fashion. The format is acceptable to multiple regulatory authorities.

Modifications to the overall application are limited to the local filing application requirements, not to the data itself. This provision speeds the overall cycle time for the application process because sponsors do not need to rewrite the data in multiple formats.

Within the CTD format, additional documents that summarize the data and its conclusions are also required. These documents are for groups of data, quality, nonclinical, and clinical. These are entitled overviews and their purpose is to provide the application reviewer with sufficient information from each section, emphasizing key parameters of the product, and justification in cases where guidance was not followed. These overarching documents were not explicit requirements of the traditional U.S.-based new drug application (NDA) format.

In addition, there are requirements for the CTD in an electronic format. The file structure and overall format of the document are strictly described in guidance documents. The applicant may not make any modifications to the structure. If the applicant has no data to report in the allocated section, it is left intact (in the formatting sequence), with a notation that ''no data exists'' for the section. The purpose of the structure is to facilitate the regulatory review and communication process between the regulatory authority and the sponsor.

The advantage to the reviewer is the transparency of the applicant's data. Reviewers can explore the body of knowledge from their desktops, compare groups of data with the touch of a button, and understand the logic of the conclusions by reviewing the overviews. Reviewers are no longer limited to sifting through piles of paper, trying to determine the relationship of a single report with the overall body of work, or postulating on how the sponsor drew the overall conclusions.

The advantage to the field inspectors is that they too are granted access to all of the applicant's information. While this was true in the past paper process, it was not practical to expect that field inspectors would be able to review all the data with the same ability as the center reviewers. Field inspectors had more limits on their time and greater breadth of workload. With the introduction of the eCTD, field inspectors can now review at will any data within the application from their desktops. They can even bring the data with them to an inspection, in its electronic format. This facility translates into better-prepared inspectors.

Thus, the prior strategy of ''assure most important is most available'' is rendered obsolete when the inspector is granted ''all access.'' A new strategy based on the full disclosure of all data needs to be adopted.

THE SYSTEMS-BASED INSPECTION

The overall purpose of a regulatory inspection is to assure the authenticity and accuracy of the data. This assurance applies to all the data in the application. Multiple ways exist to accomplishing this task. The most simplistic one for the inspectors is to identify a portion of the documents included in the application, and then request for it to be produced (as the original document) during inspection. They then make a comparison between the original document and the image (photocopy or electronic) for equality. If this comparison proves true, the assumption is that the balance of the application is also supported.

In a systems-based inspection, the inspectors may select the same documents and compare these with the originals, and continue the inspection into how each of the documents was generated. They then evaluate the "how" system (or process) for its ability to produce authentic and accurate data. They also evaluate the system for its ability to consistently produce the results presented in the application.

In general, if the inspectors find the system to have a deficiency in its ability to produce accurate and authentic results in a consistent manner, then the value of the data within the system comes under question. If multiple systems are evaluated and one or more inconsistencies are found within and between systems, then the entire application's integrity can be called into question.

The focus of these system inspections shifts the emphasis from "checking" (or reactive type of inspection) that the original data "matches" what is presented in the application, to evaluating the process by which the data is generated and analyzed (or proactive type of inspection). The basis for evaluation then becomes "if the system has integrity, then the data has integrity."

LINKING THE APPLICATION PROCESS AND THE INSPECTION PROCESS

The advances in the application process, including the rigid file structure and the ability to hyperlink and key word search in a virtual medium, have allowed the reviewers a more complete access to data. The additions that the CTD formats have added, especially the requirement for overview documents, have allowed the sponsors to better define the criteria from which they have drawn the conclusion that the data support the conclusions in the application. These additions have closed the gap between the regulator's understanding of the data and the sponsor's intent.

All these improvements are designed to reduce the overall cycle time for application approval. This time reduction then allows for the field inspectors, who are under the direction of the center reviewers, to direct the limited time they have at the site on a system, or number of systems, during the inspection. Instead of attempting to target everything, the inspection can now be more focused on a portion of the data, with an overall understanding of the process that leads to the data.

The types of data that require evaluation during an inspection include the following:

- Biobatch manufacturing
- Manufacturing of drug substance
- Excipients manufacturing
- Raw materials [current good manufacturing practice (cGMP) controls]

- Raw materials (tests, methods, and specifications)
- Composition and formulation of finished dosage form
- Container/closure system
- Labeling and package controls
- Labeling and packaging materials
- Laboratory support of methods validation
- Product (cGMP) controls
- Product tests, methods and specifications
- Product stability
- Comparison of the relevant pre-approval batch(es) and proposed commercial production batches
- Facilities, personnel, equipment qualification
- Equipment specification(s)
- Packaging and labeling (cGMP controls)
- Process validation
- Reprocessing
- Ancillary facilities

Inspections will vary on the basis of the type of product in the application. For the sake of example, let's assume that the application is in the eCTD format, and is for an aseptically processed injectable liquid product. The product's indication is not important for the example.

During the inspection, the inspector could request the following:

- The executed batch record that represents the batch of drug product used in the pivotal clinical trial that is supported by the application.

From this single record, the inspector could trace a majority of the systems used to support the application. The details are explained below.

Biobatch Manufacturing

The record can be compared with the cGMP requirements for production records and reports. If a master batch record was developed prior to this batch's manufacture, the inspector can review the process of master batch record development (including the review and approval on the record itself), the issuance, execution, review, and approval of a production record. The inspector may ask to see the master signature log to verify who has made entries into the record.

The inspector may then compare the process as defined by this record with the overview document in the eCTD.

Manufacturing of Drug Substance

If the drug substance manufacture has taken place within the same facility as the drug product manufacture, the inspector may choose to have the records associated with the batch of drug substance used in the batch of drug product presented as original documents.

Excipients Manufacturing

If the excipient manufacture has taken place within the same facility as the drug product manufacture, the inspector may choose to have the records associated with the batch of drug substance used in the batch of drug product presented as original documents.

As this example is an aseptically produced injectable liquid, it is most likely that the water for injection (WFI) used to clean equipment and for batch production is made within the facility. The inspector could then call for all the WFI production records for a period of time around the date of manufacture of the drug product production. These data will give the inspector a second opportunity to review the master batch record and production batch record system.

As WFI has specific analytical and microbiological production requirements, the inspector may call for the records associated with the establishment of those requirements. This review could give the inspector the first look into the laboratory systems within the facility.

Raw Materials (cGMP Controls)

The inspector may make requests similar to the ones for drug substance and excipients. It is likely that these materials will be purchased by the facility for use in manufacturing. The inspector may ask for purchasing specifications, shipping and receiving records (using those to trace the warehousing requirements of the cGMPs), acceptance testing, physical qualification, performance testing, and validation records for the batches of raw materials used in the production of the drug product.

Raw Materials (Tests, Methods, and Specifications)

From the records procured during the raw material review, the inspector may expand the record review to include the evidence of the training, education, and experience that each of the individuals has had that qualify them for the duties they executed within the evaluation of the batch of raw material. The inspector's expectations may include the current CV, job description, and training record for all the personnel requested. The inspector will likely compare the education requirements of the job description with the credentials of the specific employee. The inspector will want to be able to trace training events documented in the training record as they relate to the job the employee signed for in the raw material record. Again, the date of the training will be compared with the actual date that the job function was performed on.

Composition and Formulation of Finished Dosage Form

The inspector may compare the details in the executed drug product production record with the conclusions in the overview documents in the eCTD.

Container/Closure System

The packaging records contained in the drug product production record can be a source of information for the inspector. Purchasing specifications, shipping and receiving records (using those to trace the warehousing requirements of the cGMPs), acceptance testing, physical qualification, and performance testing records may be requested by the inspector for the container/closure systems used in the production record.

For the example, these materials must be aseptically processed prior to use in the drug product production. The inspector may request the records associated with the system of aseptic processing to evaluate the authenticity and accuracy of the system. This review would include ascertaining analytical and micro-biological requirements and evidence of the container/closures achieving those requirements as well.

The inspector may then include a request to review the evidence of the training, education, and experience that each of the individuals has had that qualify them for the duties they executed within the evaluation of the batch of raw material. The inspector's expectations may include the current CV, job description, and training record for all the personnel requested. The inspector will likely compare the education requirements of the job description with the credentials of the specific employee. The inspector will want to be able to trace training events documented in the training record as they relate to the job the employee signed for in the raw material record. Again the date of the training will be compared with the actual date that the job function was performed on.

Labeling and Package Controls

The inspector can use the records related to the batch of labeling and packaging materials used to manufacture the drug product to evaluate the systems of label design, approval, manufacture, acceptance, rejection, warehousing, distribution, and reconciliation.

Labeling and Packaging Materials

From these records the inspector can continue to evaluate the training, education, and experience of personnel involved in any part of the labeling system.

Laboratory Support of Methods Validation

The inspector may ask to see any records that support methods validation for any of the analytical or microbiological methods used to determine the acceptability of the drug substance, drug product, raw material, container/closure, excipients, equipment, and/or facility.

Product (cGMP Controls) and Packaging and Labeling (cGMP Controls)

The inspector may ask to see any records that support the overall controls for production. As the example is for aseptic processing, this request can include acceptable cleanability levels for all parts of the operations that contribute to the microbiological quality of the drug and/or drug substance.

Product Tests, Methods, and Specifications

Any test, method, or specification can be requested during this part of the inspection. All factors that contribute to the overall quality of the drug product can be included. These factors may include sanitization, refuse collection, decontamination, and disposal.

Facilities, Personnel, and Equipment Qualification

Facilities

As the example is for an aseptically processed injectable liquid, the inspector may ask for the diagrams and drawings of the facility in which the batch was manufactured. It is likely that facility personnel will produce the most current drawings of the facility, usually entitled "as built." These "as built" dates on the drawings may be compared with the actual date of manufacture of the batch record. If the date on the drawing is newer than the date of manufacture, then the inspector will ask for past drawings that represent the facility as it was on the date of manufacture. This review will test the system of facility document maintenance and accuracy. As this example is for an aseptic facility, this review will include the building, air handling, water treatment, and waste treatment drawings.

Personnel

From the signature taken from the master signature log, the inspector may ask to see the evidence of the training, education, and experience that all of the individuals has had that qualifies them for the duties they executed within the batch record. The inspector's expectations may include the current CV, job description, and training record for all the personnel requested. The inspector will likely compare the education requirements of the job description with the credentials of the specific employee. The inspector will want to be able to trace training events documented in the training record as they relate to the job the employee signed for in the batch record. Again, the date of the training will be compared with the actual date that the job function was performed on.

Equipment

As equipment identifiers are required in the batch record, the inspector may use those to ask for purchasing specifications, shipping and receiving records (using those to trace the warehousing requirements of the cGMPs), acceptance testing, physical instillation, qualification and performance testing, and validation records.

All these records may be compared with the standard operating procedures (SOPs) that were in place at the time of the manufacturing. A request to see those editions of the SOP will evaluate the system of documentation archival and maintenance.

As this example is in an aseptic facility, the inspector may review all cleaning and sanitization requirements for facility, personnel, and equipment. This review can be as detailed as including the verification of the sanitization status of the sterile garb used by the personnel from the signature log that were in the aseptic manufacturing suite during production of the batch of drug product.

Ancillary Facilities

As this example was for an aseptic facility, the inspection can expand into any of the support systems that produce air, water, light, and power.

Additional Links

The overview summaries in the eCTD may contain flow diagrams outlining the process and potential test points along the way. This tool would help guide the inspector when verifying that the process contained in the application is the one used to produce the batch record under review.

STRATEGIES FOR MOST SUCCESSFUL INSPECTIONS

The example above illustrates how, from a single record, an inspector can gain access into a majority of systems used to support the data in a regulatory application. So, how does a sponsor get prepared for this type of inspection? The following are some suggestions, when used over the course of the development process, will be helpful in managing the inspection outcome.

The Internal Audit Plan

Most companies engaged in pharmaceutical development have a plan to review the current status of their internal systems with the regulatory expectation of the system. For global organizations, this exercise may include a review of how operations within regions (the United States, the European Union, and Japan) compare with each other. For large organizations, this internal audit plan is generally conducted with dedicated employees and may be supplemented with external subject matter experts. In smaller organizations, it may be conducted completely by external resources.

The plan includes a review of all of the systems that an inspector would likely review. The audit observations are shared with the appropriate departmental and managerial personnel and are used to drive corrective action within the system.

The more the sponsors know about the systems in place, the better they are able to explain the overall system design and how systems have evolved over time.

The External Audit Plan or Vendor Qualification

Many companies engaged in pharmaceutical development also employ a plan to review the current status of their vendors' systems with the regulatory expectation of the system. If the system evaluations at the vendor are satisfactory, the sponsor has a higher level of assurance that the vendor's materials will have no adverse impact on the sponsor's systems.

This plan can be executed with internal and/or external resources. Subject matter experts may join the audit teams to provide guidance and insight on the specifics of the system.

Subject Matter Experts

No one organization has the expertise to know all the aspects of pharmaceutical development and that is why subject matter experts are important. Many may come from academic backgrounds and have a lot of theoretical knowledge. Using these kinds of personnel during an audit sequence can challenge the sponsor to assure that the system is robust and reproducible.

Subject matter experts are also helpful when a system failure occurs. They can potentially diagnose the root cause of the failure with more speed and accuracy than most personnel.

The creation of internal subject matter experts, where the subject is the overall development plan and its execution, is a valuable use of time and energy. Personnel tend to shift over the life of a development program. The ability to collect the overall ''rationale'' for the nonclinical program, clinical study design, and drug product design is important when questions come up during the regulatory review process.

Reference tools such as a development reports or study design documents should be made accessible to sponsor personnel. These tools can help the site personal guide the inspector during the inspection to the "rationale" of why the organization performed the work as was presented in the application.

The Power of the Overview

When using the eCTD format, the overview requirement forces the sponsor to condense and clarify many years of data into a manageable format. The use of flow diagrams and/or other visual aids orient the reader (bother reviewer and inspector) to the overall operations. These kinds of tools are useful in expediting the review process.

Data Debate versus Data Discussion

The combination of the eCTD formats and the systems-based inspection shift the discussions during the inspection from debate to discussion of the data. Because the sponsor is discussing the body of work in the overview documents, the inspector has a headstart on understanding the rationale for the developmental program. Time can be better spent on discussing the outcomes of the data rather than the data itself.

Remembering the Sponsor Knows More Than the Reviewer About the Program

No matter how well prepared the inspector is, the sponsor will always know more about the drug and drug substance than the inspector. This is not arrogant knowledge, but intellectual property knowledge, and is an advantage to the sponsor.

In turn, the reviewer and/or inspector will have inherent knowledge of other applicants' efforts (which they cannot share). If the sponsors are listening carefully during the review cycle and/or inspection, concerns the reviewers and/or inspectors may have could be revealed by the questions asked.

Keeping Accurate and Timely Notes

Any opportunity to interface with the reviewers and/or inspectors is an opportunity to learn more about their overall concerns and questions. Keeping accurate and timely notes of all email, telephone, and face-to-face conversations, and reviewing them for trends and themes are other options for keeping ahead of the inspection process.

CONCLUSION

New tools in the data presentation (eCTD) and review (systems-based approach) processes have changed the landscape of the pre-approval inspection. Reviewers and inspectors are much more knowledgeable about the entire body of work. Sponsors are providing more clarity and rationale for their development programs through the use of overviews.

Pre-approval inspections are no longer data checking exercises. They are an opportunity to show off all the great work done by the development team by convincing the reviewers and inspectors that the rationale in the application is supported by accurate and authentic data. When handled well, the outcome of the inspection should be that the reviewers' only choice is to ''approve'' the drug product.

6

A cGMP Risk Assessment and Management Strategy: Guidelines for the Pre-Approval Inspection

Carmen Medina

*Precision Consultants, Inc.,
Coronado, California, U.S.A.*

INTRODUCTION

This chapter provides a comprehensive risk assessment and management strategy aimed at helping FDA-regulated firms best prepare for the pre-approval inspection (PAI). It will also briefly discuss how to make the most of the actual inspection while in progress, and design a post-inspection risk minimization plan that will secure credible and meaningful relationships with Food and Drug Administration (FDA) officials throughout the submission review period and post-approval process.

The term risk management strategy (RMS) will be used throughout the chapter and is defined as ''a strategy that is designed around the identification, (quantitative and qualitative) analysis, mitigation, monitoring, and prevention of risk.'' Another term that will be used is risk mapping (RM). Risk mapping helps a firm identify opportunities during key phases of the drug approval process where assessing and preventing risk would prove the most valuable and reap the greatest rewards. Preemptive strikes at just the right time and at critical junctures of the drug development and approval process will invariably prevent crisis during the PAI and post-approval. So much information is revealed during the clinical trial phase of drug development only to be ignored, or not thoroughly mined, which invariably leads to public health concerns post approval. We have witnessed several examples of products in recent history that have enjoyed short-lived post-approval success only to be pulled off the market because of health concerns and complications that can be traced back to the clinical trial phase.

This chapter will not discuss *how to use* or the *principles* of specific risk management tools such as the following:

- Fault tree analysis (FTA) for which there is an ISO standard
- Hazard analysis and critical control points (HACCP)
- Failure mode and effect analysis (FMEA) for which there is an ISO standard
- Failure mode effect and criticality analysis (FMECA)
- Measurement system analysis
- Calculating risk priority numbers (RPNs)
- Statistical analysis of quality data

Suffice it to say that there are a number of tools and techniques available for the evaluation of risk, and which tool is employed depends upon what is being evaluated and what metrics are being applied.

This chapter will, however, help you design a continuous improvement system through a risk-based assessment and management strategy that is compliance centered. The author has developed a risk rating approach that defines what systems need to be measured and how, along with means to convert data to information related to risk levels that will support intelligent crisis prevention decision making within any firm. More about how to establish a risk rating is addressed in the section titled, ''Pinpointing Risk: The Performance Assessment and Needs Analysis.''

Information gathered by using the RMS presented herein, along with the risk ratings that result from the compliance-centered assessment can all be integrated into a continuous improvement program or cGMP enhancement master plan (GEM Plan™), which is an extension to the firm's RSM and RM tools and will go a long way toward minimizing risks when serious compliance issues are raised during or after the FDA PAI.

Ideally, this sort of proactive, risk management strategy should be prepared by the firm months before scheduling the PAI with the district office. This GEM Plan or risk management tool would then be leveraged during the course of the inspection, and only after compliance issues are cited by the investigator that the firm needs to demonstrate that it has been responsible in assessing itself and preparing a proactive, forward-thinking improvement plan. In fact, there are many experts within FDA-regulated industries that advocate implementing RMS the day a development project is positioned for commercialization. There will be more discussion about the GEM Plan in the section titled ''Post-inspection Risk Management and Crisis Prevention Strategies.''

DESIGNING THE PRE-INSPECTION RISK MANAGEMENT STRATEGY

Before a firm can feel comfortable that the RMS they have designed is comprehensive, integrated, and far-reaching, they must first delineate all the areas of potential risk they are uniquely faced with, from product development and clinical-trial activities right up to the day(s) of the PAI and post-inspection recovery activities. RM is a prerequisite to the design of any RMS.

In an effort to make RM a useful tool applicable to pharmaceutical, medical device and biologic products seeking approval, the following phases of the product development and approval process have been kept fairly general in order to convey how RM has application in every FDA-regulated industry. Before defining the critical phases of product development and the approval process that need to be considered and mapped for potential risk, readers should be alerted to the FDA's most recent thinking on the subject of risk assessment and management.

Understanding FDA's Latest Initiative on Risk Assessment and Management

In May 2004, the FDA published three new guidances[1] on the subject of risk assessment and management during preliminary clinical trial phases and post approval. The biggest implication of the FDA's recommendations for risk assessment and management is that they expect the entire *product life cycle* to be taken into account when designing any kind of RMS.

The guidances also suggest that the FDA has determined that the onus is entirely on the industry to address every phase of the product development, approval, and post-approval process with respect to identifying and managing risk. They repeatedly recommend that FDA-regulated industries go beyond current regulatory requirements and rigorously evaluate data from clinical trials and product complaints to prevent risk to public health. They also ask that firms apprise the FDA of their risk assessment and mitigation plans, and that they do so from a product "lifecycle" perspective.

For pre-marketing risk assessment, during clinical trial activities, the agency asks that firms develop and use *Risk Minimization Action Plans* (RiskMAP[2]) to minimize known risks. Known risks are those that become apparent during investigational studies of the product. Post-approval risk management, according

---

[1]List of FDA Guidance Documents
Pre-marketing Risk Assessment Guidance
www.fda.gov/cder/guidance/57 65dft. pdf
Development and Use of Risk Minimization Action Plans Guidance
www.fda.gov/cder/guidance/5 7 66dft. pdf
Good Pharmacovigilance Practices and Pharmacoepidemiologic Assessment Guidance www.fda.gov/cder/guidance/5767dft.pdf

[2]RiskMAP-FDA's term

to these new guidances, takes the form of good *Pharmacovigilance Practices and Pharmacoepidemiologic Assessment.* The agency's recommendations do go above and beyond what current regulations require. While there are points of overlap among the three guidance documents, they all point to the implementation of useful risk minimization techniques throughout a product's life cycle. The aim is to reduce known and unknown risks associated with a product while preserving its benefit.

You may be asking yourselves if there is anything that can be done in preparation for the FDA's new risk assessment and management initiative. The answer is an unequivocal and resounding yes, and this chapter attempts to help in those preparations. The need for firms to perform a proactive gap analysis of their risk management status and safeguard all phases of activities and subsequently inform the agency of its risk management plan is what these three new guidance documents require; and while these guidances are not codified law, they certainly carry weight with respect to the agency's expectations toward establishing industry standards.

Risk assessment during product development would require that the sponsor study an adequate number of patients and that they make the number to be studied commensurate with risks associated with the investigational product. It would also require that the sponsor use the most appropriate study design, utilize serious adverse event (SAE) reporting and annual reports appropriately to capture, and subsequently mine (serious and nonserious) reported events. The FDA has asked that as part of risk assessment and management, the sponsor target suspected safety issues and thoroughly understand their impact on public health.

The agency has asked that the industry recognize that the various phases of clinical investigation are an iterative process, and as the sponsor learns more about the product, they do more regarding suspected risks. Investigational studies must be designed to run for an adequate period—long enough to have potential safety risks reveal themselves. It is essential to assess the product's risk-benefit balance and evaluate underlying risks in an effort to comprehensively characterize a safety profile.

Additionally, the FDA expects that the sponsor to quantify occurrence of expected, low-frequency adverse events (AEs) and develop a meaningful alert signal system to enable clinical investigators to become immediately aware of dangerously high occurrences. Upon becoming aware, the FDA expects the sponsor to develop and implement tools to minimize those risks. They must also periodically evaluate the effectiveness of those risk assessment and management tools and adjust as needed to enhance the risk-benefit balance.

It is important to understand the unique risk potential during the different phases of product development and accept that risk assessment during product development is indeed different from assessing risk post approval, even if some of the same tools are used.

Phase 3–Oriented Risk Assessment and Management Techniques

By Phase 3, a dose-response should be established, the efficacy confirmed, and the lowest possible dose or regimen determined, particularly when used for a chronic condition. Do not wait until the post-approval to assess potential inter-actions (concomitant use assessment); or to assess population subsets. Reserve blood samples should be collected at this time, in the event retrospective analysis is required. Phase 3 is the time when the sponsor should design expansive safety data collection systems to enable adequate data mining.

Additional Phase 3–oriented risk assessment and management activities might include kidney, bonemarrow, and liver toxicity; potential drug inter-actions; polymorphic metabolism; and immunogenicity. Risk assessment and management related to data analysis and presentation might take the form of a standardized and consistent terminology, accurate coding, installation of an audit trail, appropriate system for combining or dividing AE codes, and concisely defining syndromes.

This is the phase during which risk assessment and management should focus on medication error prevention analysis (MEPA), the detection of unanticipated interactions, ensuring proper data analysis and presentation, and the development of appropriate risk assessment tools specific to the product. Analyze temporal associations, keeping in mind that aggregate safety data tends to cloud this potential risk. Use titration studies when dose effect is a factor. Attempt to glean potential risk patterns from pooled data wherever possible. It is also important to properly reflect withdrawal and *dropout* rationale—unexplained withdrawals raise red flags with the agency.

It would also be prudent to install long-term follow-up for AEs that are received late. Another very effective risk management tool is the investigational study's case report form (CRF). It should be comprehensive enough to capture, help identify, and predict potential risks associated with the investigational product.

There are many things that can be done during Phase 3 that would allow the sponsor to identify an unusual risk profile. They can ensure product labeling and use physician education to minimize risk. The bottom line is that the sponsor must target one or more product safety concerns, and translate risk assessment into pragmatic, specific, and measurable risk management objectives.

Post-marketing Risk Management Guidance

The guidance documents also address risk assessment and management activities after the product is in commerce. The agency asks that firms identify observable safety signals; routinely assess pharmacoepidemiologic data; develop a phar-macovigilance plan using appropriate tools; and install mechanisms to capture safety data. Adequate data mining, comparison of observed rate versus expected rate of incidents and trending and correlating data, whenever possible is also expected in an effort to assess and mitigate potential risk.

Most recently, the agency has asked the industry to evaluate and track "like-moiety" products. It is expected that firms will rigorously investigate single case reports of what are considered rare events, and that only qualified professionals, such as licensed pharmacists and nurses monitor reportable events. Also, where warranted, firms should conduct formal Phase 4 studies if potential risk is suspected.

RISK MAPPING: AN EFFECTIVE RISK ASSESSMENT AND MANAGEMENT TOOL

With the FDA's current risk assessment and management expectations and recommendations as a backdrop, we can return to the technique of RM and how it can be applied to any phase of product development, the review and approval process, and commercialization. The various critical phases of the product development and approval process are generally presented below to convey how RM has application:

- Development or design controls
- Regulatory submission preparation
- Manufacture of clinical trial material (CTM) or device first article
- Clinical trial activities (Phases 1, 2, and 3)
- Iterative process between FDA submission reviewers and the sponsor
- Scale-up studies
- Facilities and equipment qualification
- Process validation
- Pre-approval inspection
- Post-inspection recovery activities
- Phase 4 clinical studies

Table 1 represents a sample of potential risks associated with each critical stage along the product development and approval process and what could be done to mitigate that risk, once it has been identified as a possibility within the context of a specific product and given a project's particular set of circumstances. There are many more risks that can be delineated for each of these stages; however, while every PAI project faces similar risks, RM must be *project specific* for it to be meaningful.

At any one of these critical stages potential risk exists, the extent to which it exists can be assessed, managed, and, in many cases, mitigated. Potential landmines along the trail of any of these critical stages can be mapped and either preempted or adequately managed to prevent major impediments to a successful PAI.

Note that all of the risks delineated are quite general—a more concise map can be developed once a full assessment of the overall project is performed. Another type of assessment that can be performed, along with this type of RM is a gap analysis in which all aspects of the overall operation and project quality are evaluated, and upon which the subsequent needs analysis is based.

Table 1 Critical Stages Along the Product Development and Approval Process

Stage	Potential risk	Mitigation
Development	Inadequate/unreliable supplier of API	Install quality organization capable of external audits and with access to upper management for business relationship management.
	Premature process definition and/or specification setting	Timely and periodic development report review; adequate time/resources for development activities
	Inadequate test method development	Early involvement of QC subject matter experts and use of recognized consensus standards
	Drug product formulation is inadequate or inappropriate	Ensure project teams include marketing, regulatory, pharm/tox[a], and clinical are involved early in development process.
Regulatory submission preparation	Faulty or sloppy data	Build regulatory involvement throughout product development; QA/QC reviews and verifications integrated throughout documentation process
		Install adequate document control system to ensure versioning; also facilitates interdepartmental review
		Use risk-based approach to test method development and validation
Manufacture of CTM	Lack of cGMPs where needed	Install rigorous QA oversight throughout organization integrated with regulatory function to ensure appropriate ramp-up
	Material inadequate for trial	Ensure preliminary analytical testing in place and initial data supports trial logistics (stability, container/closure, etc.)
	Qualification and validation deficiencies	Employ risk-based validation approach

(Continued)

Table 1 *(Continued)*

Stage	Potential risk	Mitigation
Clinical trial activities (Phases 1, 2, and 3)	Poor quality oversight	Ensure iterative process between FDA's submission reviewers and the sponsor; install QA check points throughout the trial; qualify CRO and monitor
	Poor SAE tracking	Install regulatory oversight and systems implementation/ training; entire clinical monitoring effort managed to ensure rapid information retrieval, cleaning, and analysis
	Poor communication	Design and install information interfaces to ensure timely organizational communication; formalize information transfer activities
	Computer systems not adequately validated	Use risk-based validation approach
Scale-up studies	Poor DOE; inadequate initial process characterization	Employ adequate statistical support during development, utilize process mapping, and process analysis tools
	Facilities not ready	Ensure adequate project management, budget, and business processes support realistic development plan.
Facilities and equipment qualification	Old equipment and facility	Understand and prepare for process constraints, likely failure modes and critical parameters; stay current on industry standards and regulatory expectations; plan realistic budget and organizational financial resources, utilize predictive maintenance techniques
Process validation	Not able to validate; poor initial process characterization, worst cases conditions not defined and challenged	Ensure manufacturing is linked to developmental work; understand critical parameters to consistent product quality, utilize DOE approach

(Continued)

Table 1 *(Continued)*

Stage	Potential risk	Mitigation
PAI	Not prepared	Ensure adequate PAI training for all involved organizational units; use expert project manager; assign PAI leaders for pertinent product and quality processes
	Not understanding requirements	Hire experts on what is required; plan for success
Post-inspection recovery activities	Incomplete response to investigator's concerns	Be prepared to address verbally delivered findings when raised or at exit presentation; prepare a GEM Plan
	Inadequate communication to stakeholders	Gain corporate support for teams to address 483 findings; ensure EIR distributed for action item assignment and tracking; formalize involvement of project management resources
Phase 4 clinical studies	Does not address clinical concerns or epidemiological trends	Design studies that target appropriate sub-groups; Assess & mine pharmaco-epidemiologic data; Develop a pharmaco-vigilance plan

[a]pharmacology and toxicology.
Abbreviations: API, active pharmaceutical ingredient; QC, quality control; QA, quality assurance; cGMPs, current good manufacturing practices; CTM, clinical trial material; FDA, Food and Drug Administration; CRO, contract research organization; SAE, serious adverse event, DOE, design of experiment, PAI, pre-approval inspection; GEM Plan, GMP enhancement master plan; EIR, establishment inspection report.

PINPOINTING RISK: THE PERFORMANCE ASSESSMENT AND NEEDS ANALYSIS

Third-party consultants are frequently retained to conduct a gap analysis or performance assessment of an organization's overall operation. A performance assessment is an effective mechanism for evaluating all aspects of a company's operation, but it does not necessarily disclose what will be needed to install meaningful and sustainable improvements. A needs analysis should follow on the heels of a performance assessment in order to ensure that the genesis of the deficiency or performance failure has been properly identified. A performance assessment should observe specific quality attributes as well as perform data-driven analysis. For example, a measure (frequency or comparative rate) of the number of invalidated out-of-specification (OOS) results generated in a laboratory might help in the design of an innovative training program targeting the laboratory personnel generating those results. A needs analysis will illuminate the real cause of the performance failure or deficiency, leading to genuine resolution. Without pinpointing the cause beneath the problem, resources might be wasted on what is suspected, rather than on what is needed.

For example, personnel performance that falls short may not have anything to do with lack of knowledge, training, or training materials; it may have more to do with lack of motivation or inadequate incentives. While the performance assessment is the study that delineates the actual quality and compliance status of all pertinent aspects of the operation in question—the needs analysis offers possible solutions on the basis of coordinated qualitative and quantitative, data-driven information to determine exactly what is needed to reverse the performance issues.

It is quite helpful when performance outputs are measurable, for example, fewer manufacturing deviations, rejects, reworks, OOS results, etc. If a performance assessment shows that what the organization is lacking is effective training programs, the obvious solution would be to design innovative tools, job aids, coaching, electronic performance support, documents, and training sessions that will enhance performance through training. However, the needs analysis might show that it is lack of proper incentives that most affects performance, in which case the solution would focus on creative ways to increase employee incentives in an effort to enhance performance. Alternatively, the study might show that the solution is a cross-functional one—the point is that the organization cannot effectively address personnel performance, if it is not certain of the various factors that drive it.

The performance assessment coupled with a needs analysis provides an opportunity to look deep within the company, its personnel, its practices, and management to determine what the underlying weaknesses truly are, and, subsequently, design the most appropriate interventions. Such comprehensive assessment and analysis rely upon various sources of information, everything from personnel interviews and record reviews to assessments of products and work output trends. These can serve to reveal roadblocks, as well as best practices. Together, performance assessment and needs analysis serve as a springboard for risk assessment and management, as well as continuous improvement.

Assessment of Critical Compliance and Quality Systems

There are a number of indispensable quality and compliance systems required by medical device, pharmaceutical, and biologics regulations without which a firm simply could not operate, and that also serve as meaningful performance indicators when performing a gap analysis or performance assessment. Table 2 lists several critical quality/compliance systems.

While recall prevention is not a codified regulatory requirement for any FDA-regulated industry, it certainly is an industry standard and FDA expectation that firms install whatever measures are needed to protect the public health, and place only acceptable material and product into commerce.

Table 2 Essential Quality and Compliance Systems

---

1. Process validation
2. Stability testing program
3. Computer validation
4. Supplier qualification
5. Raw material and component qualification
6. Laboratory accountability and traceability
7. Analytical methods validation
8. Personnel training
9. Handling OOS and OOT results
10. Handling manufacturing deviations
11. Corrective and preventative actions
12. Statistical rationale for sampling methodology
13. Proactive QA program
14. Internal material/product certification
15. QA review and approval prior to release
16. Calibration and PM program
17. Master and production record development
18. Change management system
19. Consumer complaint handling system
20. Environment monitoring/contamination control
21. Regulatory tracking
22. Facility, equipment, and support utilities qualification
23. Cleaning validation
24. Scale-up program
25. Recall prevention program[a]

---

[a]Recall prevention is not a regulatory requirement.
Abbreviations: OOS, out of specification; OOT, out of trend; QA,
quality assurance; PM, preventive maintenance.

Under the auspices of a performance assessment, gap analysis, or internal audit, a thorough qualitative and quantitative assessment of the quality systems listed in Table 2 would provide significant insight into both the compliance status of the overall operation and the risk level facing the company.

The following compliance rating system and criteria can be applied to the overall operation of any FDA-regulated company as a result of a comprehensive performance assessment, and would form the basis for designing a compliance-centered risk assessment and management plan. There are many other aspects to a business that impact potential risk, such as accounting and executive management practices, but the emphasis of this chapter is preparing for the PAI by using RMS; additionally, when discussing risk, regulatory compliance weighs heavily upon the overall risk and liability facing FDA-regulated companies.

Table 3 Performance Assessment of 24 Individual Quality Systems' Inspection
Readiness

Is the quality system formalized?[a]	Does the formalized, quality system require minimal or extensive enhancement?
No (some % of 24)	Not applicable
Yes (some % of 24)	Yes (some %) Some % of the number of formalized systems need some degree of enhancement.

[a]Formalized (i.e., system is rendered into an approved procedural format, trained to and implemented).

Possible Compliance Rating System

Let us hypothetically elect to use a number system that ranks the state of regulatory compliance within a company by first individually evaluating the essential quality systems in Table 2 in order to determine regulatory consistency, along with the level of compliance to which each system has been designed and implemented within the company. After establishing the level of compliance for each individual quality system, that information is evaluated collectively in order to assign an overall compliance rating. The first step could be depicted as show in Table 3.

Formalization of the 24 systems referenced in Table 2 is a regulatory requirement for pharmaceutical (R_x) manufacturing and cGMP compliance—recall prevention was not counted. Let us imagine for a moment that of these 24 required systems, 12 do not currently exist at the firm. These systems directly relate to inspection readiness and application integrity and acceptability during submission preparation. Additionally, the remaining 12 systems that actually do exist would be deemed deficient to some degree, during an FDA inspection, since they do not contain all of the essential compliance components required by the regulations, and minor or major enhancement would be needed to bring them into full consistency with the law, industry standards, and FDA expectations.

Now Table 3 would actually be represented as in Table 4.

As a result of these compliance-centered findings, the company would be deemed "not inspection-ready" and would receive a compliance rating of 2 if the following criteria in Table 5 were applied.

This hypothetical company falls into the "high risk" quadrant, as depicted in Figure 1. Risks associated with compliance ratings 1–4, could be as defined in the figure.

A series of questions and decisions would naturally follow as a result of the information garnered from this extensive assessment and predicted risk:

- What is the total liability facing the company: financially, public credibility, etc.?
- What degree of financial loss can result from this level of risk?

- What degree of financial risk can the company afford to sustain, if any?
- What degree of any kind of risk can the company withstand?
- What can be done to mitigate the various risks facing the company?
- What can the company do to preempt crisis in light of the risk it is facing?
- What is the company policy on crisis prevention?

And so on. The questions that arise from this degree of detailed qualitative and quantitative information are virtually endless, and the actions that must be taken in response to this critical mass of information equally substantial.

Table 4 Performance Assessment of 24 Individual Quality Systems' Inspection Readiness

Is the quality system formalized?[a]	Does the formalized, quality system require minimal or extensive enhancement?
No (50%)	Not applicable
Yes (50%)	Yes (100%) 100% of the number of formalized systems need some degree of regulatory enhancement.

[a]Formalized (i.e., system is rendered into an approved procedural format, trained to and implemented).

Table 5 Compliance Rating System: Levels 1, 2, 3, and 4

1. Represents an operation that incorporates less than 50% of the essential elements required by the cGMPs, other applicable regulations (ICH guidelines), current industry standards, and FDA expectations for this type of company.
2. Represents an operation that incorporates some (more than 50%) of the essential elements required by the cGMPs, other applicable regulations, current industry standards, and FDA expectations for this type of company, but for which additional formalization and enhancement is still needed for the company to be fully compliant and sheltered from regulatory liabilities. This organization does not have a formalized, continuous improvement plan in place.
3. Represents an operation that incorporates all of the essential elements required by the cGMPs, other applicable regulations, current industry standards, and FDA expectations for this type of company, but for which personnel adherence levels need improvement for the company to be fully compliant. This organization has a formalized, continuous improvement plan in place.
4. Represents an operation that incorporates 100% of the essential elements required by the cGMPs or applicable regulations, current industry standards, and FDA expectations for this type of company and has achieved a high level of personnel compliance to internal procedures and FDA regulations. This organization has a formalized, continuous improvement plan underway.

Abbreviations: cGMPs, current good manufacturing practices; ICH, International Conference on Harmonization; FDA, Food and Drug Administration.

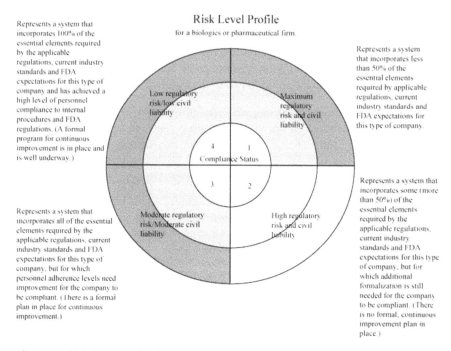

Represents a system that incorporates 100% of the essential elements required by the applicable regulations, current industry standards and FDA expectations for this type of company and has achieved a high level of personnel compliance to internal procedures and FDA regulations. (A formal program for continuous improvement is in place and is well underway.)

Represents a system that incorporates less than 50% of the essential elements required by applicable regulations, current industry standards and FDA expectations for this type of company.

Represents a system that incorporates all of the essential elements required by the applicable regulations, current industry standards and FDA expectations for this type of company, but for which personnel adherence levels need improvement for the company to be compliant. (There is a formal plan in place for continuous improvement.)

Represents a system that incorporates some (more than 50%) of the essential elements required by the applicable regulations, current industry standards and FDA expectations for this type of company, but for which additional formalization is still needed for the company to be compliant. (There is no formal, continuous improvement plan in place.)

Figure 1   Risk level profile for a biologics or pharmaceutical firm.

When thinking in terms of risk assessment and management, it is important to consider a number of factors such as the size and financial status of the company; the degree of public trust that company has earned and could stand to lose; the degree to which these findings impact just one product or all company products; whether this compromises a favorable regulatory history with the FDA, and much more.

Once the performance assessment phase of the risk assessment methodology (RAM) strategy has led to the appropriate risk status or rating, the only way to mitigate any of the above concerns would be to develop a risk management plan and ensure that the executive management will invest the needed resources to fully implement the plan.

We have discussed RAM from the perspective of RM throughout critical stages of the product development and approval process, as well as using a compliance-centered approach to evaluate overall compliance and quality performance within a company in an effort to arrive at an understanding of the liabilities or degree of risk facing the company. Assessment without the appropriate follow-up management plan is more harmful than no assessment at all.

It would be quite disheartening to an organization's personnel, should they become aware that the executive management elected to carry out a comprehensive assessment only to abort the risk management and crisis prevention phase. There is also a large measure of negligence that can be assigned to company management that launches an assessment of such magnitude, and elects not to move forward with the risk management phase, or who develops a compliance master plan for the sole purpose of having a defensive tool should a PAI inspection turn violative. As we mentioned earlier, a RAM strategy comprises both assessment and management. The GEM Plan is an effective and powerful tool in aiding the company's management should a PAI become challenging; however, once the management has become aware of prevailing issues that may surface during the inspection, they must act responsibly and address the issues and not wait for the FDA to impose a mandatory corrective action plan.

The GEM Plan is the proverbial *"ace up the sleeve"* and, since it exposes areas for improvement that may not have been uncovered during the inspection, should only be used after the firm becomes aware of the investigator's intention to issue an FD-483 and the specific nature and extent of the citations. While the GEM Plan is a valuable countermeasure during an inspection, it should never be used solely as a defensive tool. Remediation should immediately follow the assessment phase of any GEM Plan initiative.

RISK MANAGEMENT DURING ALL PHASES OF CLINICAL TRIAL ACTIVITIES

There are a number of things that can be done to evaluate and manage risks during the investigation study phase of the product development and approval process, some of which the FDA delineated in their latest risk management guidances distilled above. Let us examine the types of quality assurance (QA) issues that could seriously assault data integrity and discuss how to prepare and manage risks during the various clinical investigation phases.

Compliance Factors Related to Clinical Trial Activities

Global harmonization efforts, addressing various compliance and quality areas like stability, clinical trials, and validation of analytical methods have been underway for over a decade. These worldwide efforts have resulted in the publication of several internationally accepted guidelines related to good clinical practices (GCPs). The International Committee for Harmonization (ICH) has marshaled the publication of these guidelines. Clinical trial activities have tripled over the past 10 years, and the growth of contract research and site management organizations (CROs, SMOs, respectively) has responded to the growing need for clinical trial design, statistical analysis, and trial QA.

Coupled with this increased demand for third-party support of clinical trial activities, is a heightened scrutiny from the FDA and other worldwide regulatory agencies over sponsor-related activities. The number of FD-483 citations, warning letters, and trial disqualifications related to noncompliance during clinical activities suggests that FDA's increased sensitivity is not unwarranted.

Headquarters will closely evaluate a firm's clinical trial activities before the Office of Regulatory Affairs (ORA) is alerted to schedule a PAI of the firm. Not long ago, the FDA instituted a special investigations unit to monitor and conduct inspections of biomedical research associated with pending applications. As a result, it has become routine to have the activities related to the conduct of clinical studies and the manufacture of CTM scrutinized by the agency shortly after the investigational new drug (IND) application is made. There is a chance that a clinical trial inspection may occur after the new drug application (NDA) has been submitted and the PAI date been set; but it is more likely to occur during later phases of development, such as Phases 2 and 3. As such, it is essential that firms assess potential liabilities during their clinical trial activities before the FDA sets about doing so. A typical risk related to noncompliance during clinical studies is the FDA imposing a ''clinical-hold'' resulting in significant delays in product approval. One way to minimize risk is to ensure that a performance assessment of study conduct and CTM manufacture occurs prior to scheduling the PAI. Here are some of the highly visible investigational activities that must be evaluated and secured prior to scheduling the PAI.

Contract Research Organizations (CROs)

An institution or a pharmaceutical, medical device or biotech company, herein referred to as the sponsor, typically contracts a CRO. They are typically contracted to manage some or all aspects related to a clinical trial. The FDA's expectations related to the sponsor's responsibilities to adequately evaluate and qualify the CRO are no different from any other supplier qualification requirements. CROs selected to manage a sponsor's investigational study activities must have a concise understanding of their QA responsibilities. While the sponsor may transfer any or all sponsor-related responsibilities to the CRO (provided the transfer of duties is delineated in a formal agreement), the ultimate responsibility for the quality and integrity of the investigational study resides with the sponsor. For a number of reasons, the FDA has become more interested in assessing the QA and risk assessment measures exacted by CROs and sponsors during clinical trial activities. One compelling reason, in addition to the increase in clinical trials and corresponding violative inspectional outcomes, are the high profile products that have gained FDA approval only to be subsequently withdrawn from commerce because of legitimate public health concerns.

The sponsor and CRO must formally define their respective roles and responsibilities in a legally binding contract. A well-defined interface between the sponsor and the CRO can serve as a viable risk management tool in that it establishes the compliance and quality standards to be applied during the course of the investigational study. It behooves the sponsor to ensure that CROs have their own QA infrastructure, even if they are asked to employ some of the sponsor's specific standard operating procedures (SOPs) for certain aspects of the clinical trial.

Table 6 delineates some of the general operating procedures that the CRO ought to have in place prior to the sponsor contracting that organization.

Should the sponsor have SOPs for similar activities, a determination must be made early on as to which set of procedures will be employed.

Table 6 Essential CRO Procedures

1. Personnel training
2. Training documentation and records
3. Change management
4. Internal audit program
5. Complaint handling and follow-up
6. Software validation
7. Quality systems
8. Documentation system
9. SOPs, work instructions, policies
10. Regulatory file maintenance
11. Study master file table of contents
12. Protocol/amendment development and maintenance system
13. Investigator selection system
14. Study start-up activities including site and investigator training
15. Confidentiality agreements
16. Patient consent log
17. Laboratory certification program
18. Investigators and subinvestigator resume maintenance
19. Ethics committee membership lists
20. Ethics committee approval letters
21. Approved patient consent forms
22. Safety data (complaints: ADRs and SAEs)
23. Service contracts
24. Correspondence to ethics committee (progress and site reports)
25. Site qualifications and monitoring visits and reports
26. Regulatory approved labeling and inserts
27. Investigator brochures
28. Updates to regulatory documents
29. Drug accountability forms (dispensing and reconciliation)
30. Drug supplies and security
31. cGCP regulations training program
32. Inclusion/exclusion criteria compliance system
33. Review of site status: issues and resolution
34. CRF monitoring to source data
35. Handling of reported/unreported complaints
36. Qualification of IRB
37. Handling of investigator compliance with study protocol

(Continued)

Table 6 *(Continued)*

38.	Enrollment log; activities and investigator incentives
39.	Compliance with new HIPAA[a] confidentiality
40.	Third-party subcontracting trial activities
41.	Protocol deviations with appropriate follow-up
42.	Enrollment activities

[a]HIPAA refers to "Standards for Privacy of Individually Identifiable Health Information" found in 45 CFR Parts 160 and 164.
Abbreviations: CRO, contract research organizations; SOPs, standard operating procedures; ADRs, adverse drug reactions; SAEs, serious adverse event; cGCP, current good clinical practices; CRF, case report forms; IRB, institutional review boards.

Another area that should be assessed for potential risk is "investigator compensation" for participation in the investigational study. Ensure that only appropriate incentives are offered to investigators for high and rapid patient enrollment.

Training of investigators and subinvestigators is another potential risk area that can tremendously impact the investigator's compliance with investigational protocol requirements, and, ultimately, the success of the trial.

The following is a brief summary of some of the many areas that become potential liabilities during clinical trial activities, and could result in the disqualification of the study.

Within the Investigator Sites

- Lack of subinvestigator training
- Inappropriate storage of CTM
- Inadequate facilities—too cramped, not private or secure
- Too many trials going on at once

Labeling of CTM

- Does not comply with protocol requirements with respect to
  - secure storage of labels,
  - accurate issuance,
  - adequate transport,
  - precise counting and reconciliation, and
  - zero defects for line clearance.

Packaging of CTM

- Does not support investigational protocol.
- Package design does not conform with dispensing needs and stability requirements.

- Does not display storage instructions on package.
- Does not exercise caution when packaging for multiple protocols.
- Does not ensure appropriate coding for blinded studies.

Handling of CTM Between Sponsor and Investigator Sites

- Is not carefully monitored.
- Manipulations of the CTM, such as changing labels and packaging, or adjusting retest dates, are not left to the sponsor or manufacturer.

Distribution of CTM

- Avoids traceability within investigator sites and everywhere else.

Stability Testing of CTM

- No retest dates on all material undergoing stability testing.
- No stability study for all CTM.
- Inadequate qualification of contracted services (manufacturing and laboratories).

Shipping/Storage Validation of CTM

- Does not maintain CTM integrity throughout the duration of the trial.
- Is not supported with a formal shipping and storage stability study
- Lacks reconciliation and accountability for all CTM in the field.

Quality and Medical Complaints Related to Use of CTM

- If safety reporting for all SAEs are not made immediately to the sponsor, unless exempted by the protocol.
- If AEs and deaths are not reported within proscribed regulatory time frames.
- If technical or quality complaints related to CTM are not adequately handled and followed up by the manufacturer and sponsor.

Quality Assurance and Oversight During Clinical Trial Activities

All activities during the investigational study phase require some form of QA oversight and risk assessment and management. The extent to which a CRO and a sponsor invest in assuring quality through risk-based approaches depends on the size and nature of the trial, the sort of protocol and type of product, and the breadth of their QA commitment.

CRFs and source documents should be routinely reviewed and authenticated. Data reported on the CRF, typically derived from source documents, must be consistent with source documents and any discrepancies explained. The sponsor or the CRO must ensure that adverse drug reactions are reported to investigators, institutional review boards (IRBs), and regulatory authorities within required time frames; this would include serious and unexpected drug reactions. Ongoing safety, information, and evaluation of the CTM by the sponsor must be provided throughout the duration of the trial. The CRO or the sponsor's QA staff must periodically audit trip-monitoring reports.

Investigator Noncompliance

Another area that has led to the unfortunate disqualification of many trials is nonconformance to the investigational protocol by medical investigators and subinvestigators. The sponsor or the CRO is responsible for training the investigators to the protocol, and ensuring they comprehend their responsibilities to the FDA and the investigational plan. The FDA expects that where noncompliance to investigational protocols, SOPs, cGCPs, and FDA guidelines occurs, the sponsor or the CRO will secure compliance.

Additionally, if field monitoring or QA staff identify serious or persistent noncompliance on behalf of an investigator, the sponsor or the CRO is expected to terminate the investigator's participation in the trial. The FDA should also be notified of such actions.

Change control is tremendously important during the manufacture of CTM for a number of reasons; however, since the formulation may not be nailed down during Phase 2 or early Phase 3, it is particularly important to track changes.

Risk Management During Clinical Trials: The Change Control System

Some risks typically encountered during Phases 2 and 3 that the sponsor should be on the lookout for:

- Clinically tested versus formulation to be marketed
- Dosage different from marketed version
- Reformulation of the marketed version
- Changes in manufacturing process
- Change in manufacturing site
- Change in analytical methods
- Change in route of administration without new studies to support efficacy
- Changes to established cleaning validation studies
- ADEs that lead to labeling changes
- Suspected adverse reactions (SARs) that lead to investigator brochure and labeling changes

Additionally, inadequately managed changes to the following elements, during a trial, could invalidate the results.

- Protocol revisions
- Consent form

- CRO infrastructure (lack of QA)
- Statistical analysis methods
- Premature termination of trial
- Trial monitoring frequency
- IRB constitution
- Pivotal personnel turnover at
  - Investigator sites
  - CRO
  - Sponsor

Risk Management During the Manufacture of Clinical Trial Materials

While this article necessarily limits extensive coverage of the compliance issues related to CTM manufacturing, a brief discussion elucidating the various aspects of CTM manufacture that must be considered when designing of a comprehensive risk assessment and management strategy during clinical trial phases is warranted.

First and foremost, CTM manufacture must apply cGMPs, where appropriate according to 501(a)(2)(B), which requires all drug products be manufactured in accordance with cGMP. The preamble to the 1978 cGMP regulations states that cGMP regulations are applicable to the preparation of any drug product for administration to humans or animals. Additionally, there is a 1991 guideline for preparation of IND products. While it does not cover all manufacturing situations, or fully address, FDA's expectation that an incremental approach to cGMP compliance is acceptable for the manufacture of CTM—a great deal of what the guidance does require has been distilled below.

Draft FDA guidance document for Phase 1 IND and complementing regulations articulate the agency's intent to appropriately implement an incremental approach to cGMP compliance for clinical investigational products, and recognizes that the extent of cGMP implementation may differ between investigational and commercial manufacturing, and throughout the various phases of clinical study. This guidance applies to both pharmaceutical and biological drug products.

It urges sponsors to utilize appropriate quality control (QC) standards such as well-defined procedures, adequately controlled equipment, and accurate recording of all data. In an effort to reduce risk, sponsors should use available technology and resources to facilitate product development such as disposable equipment and process aids, prepackaged water for injection, and sterilization contract manufacturing and testing facilities wherever possible.

A great risk often presented during both commercial and CTM manufacturing is cross-contamination. Sponsors can reduce the risk of contamination and cross-contamination through adequate evaluation of production environments, identification of potential hazards such as chemicals and adventitious agents from previous or concurrent production, and ensuring that previous agents are appropriately removed. Additionally, personnel must have the appropriate education, experience, and training to perform assigned functions. QC functions must be established for every manufacturer of CTM, and their responsibilities must be documented in writing and include examination of components, containers, closures, in-process materials, packaging and labeling materials review, approval of production and testing procedures, and acceptance criteria.

The review of completed production, packaging, and labeling records for the release or rejection of each clinical batch is the responsibility of QA personnel.

Facilities used for the manufacture of CTM must have adequate and clearly demarcated work areas for intended tasks. Water must be of appropriate source and quality, and the area must employ adequate air handling to prevent contamination and cross-contamination. The equipment used must be in proper working condition, maintained and calibrated, cleaned and sanitized at appropriate intervals, and suitable for its intended use. Equipment must not be reactive, additive, or absorptive with CTM being made, and production staff must identify and document equipment and utensils used in production records. Similarly, the control of components requires written procedures describing handling and control of components, a certificate of analysis or other documentation for components to ensure conformance with specified attributes, records of receipt, quantity, supplier's name, lot number, and corresponding expiration dates. Production documents must include laboratory and production data detailing components, equipment and procedures used, and account of all changes in processes and procedures. Change control is an essential quality system during the manufacture of CTM.

The controls for laboratory activities related to CTM manufacture require formalized procedures under controlled conditions, scientifically sound analytical procedures, and properly calibrated and maintained analytical equipment and apparatus. CTM requires formal study of the material's stability to support use of product throughout the entire investigational period.

Container closure and labeling of CTM is generally very strict, and requires the packaging to protect the material from alteration, contamination, and damage during storage, handling, and shipping. Labeling practices must aim to prevent material mix-ups. Distribution of CTM requires that the transport be described from the point of production to the investigational site where the subject will receive it.

CTM–record keeping requires that all records related to manufacture, quality control, and shipping are retained for at least two years after approval of marketing application, or until two years after shipment and delivery of the product is discontinued and the FDA is notified.

The application of cGMPs controls to screening IND and microdose studies should be proportional to the scale and scope of the operation.

An area or room can be used for multiple purposes and products, provided that only one product is produced in an area at any given time and that appropriate cleaning and changeover procedures are employed to ensure no carryover of materials occurs.

Some production systems may warrant additional safeguards to protect personnel from pathogenic microorganisms such as spore-forming microorganisms, live viral vaccines, and gene therapy vectors.

Investigational products intended to be sterile require specific precautions such as adequate personnel training in aseptic techniques. Additionally, aseptic manipulation should be conducted in a class 100 environment. Alternatively, a laminar flow hood can be used to assure appropriate air quality of an aseptic environment.

During the Inspection Process: Risk Assessment and Management Techniques

In an effort not to rehash what so many insightful authors before me have written about managing and controlling the FDA inspection process, this section will focus on how to assess, manage, and mitigate risk during the PAI.

There are many ways to stay on top of, even ahead of, the investigator(s) during their inspection of an operation.

From the moment the investigator(s) arrive, to the time they leave, everyone has a role to play and unique responsibilities to carry out. The challenges of reducing the risks associated with the PAI process can be significantly mitigated by making the firm's involvement a synchronized endeavor, well rehearsed, and carefully planned with nothing left to chance. There are many crucial elements that must come together during the inspection if it is to end favorably, whether or not a List of Observations (FD-483 form) is issued. This is the time when the firm's risk assessment and management skills will play a crucial role in preventing crisis.

One way to manage this risky process known as the PAI is to have performed a comprehensive performance assessment and RM throughout product development, scale-up, technology transfer, and certainly prior to submitting the formal application to the agency. Rehearsing the inspection process in a mocked FDA-style is always very useful, particularly if personnel experienced with the FDA inspection process take part in the exercise. It is also useful to contract a former FDA consultant with experience performing field inspections to lead the group in this exercise and provide meaningful feedback regarding group dynamics, best practices, liabilities, and areas for improvement.

Another indispensable tool for managing this challenging event is the company policy or SOP related to the FDA inspection process. It should be concisely delineated and address all aspects of the inspection process from who will act as the inspection liaison and what minutes and reports will be generated to who will be notified, who the spokesperson will be, and what post-inspection actions must be taken.

Prior to the conclusion of the inspection, and on a daily basis, the firm must insist upon debriefing sessions with the investigator(s) in an effort to gauge the course of the inspection and learn what issues and concerns remain in the investigator's mind. These sessions are conducted under the management's supervision, and focus primarily upon the investigator's perspective of the day's events. It is appropriate to inquire if there are any outstanding issues or concerns; additional information requests, clarification needs—all on an effort to assess risk.

It is important to gauge the investigator's sensitivities by asking:

- What time they intend to return tomorrow?
- What might they be interested in examining?
- What might the firm prepare for the next day in an effort to move things along?
- What is the additional information, if any, needed to augment what was already reviewed?

Their answers to these questions will give the firm insight into the concerns and questions that may have arisen in the investigator's mind during that day's inspection activities. This is a direct risk assessment tool that can be used throughout the inspection process.

All potential risks, concerns, and questions must be raised and addressed on a daily basis. The exit conference is not the time to learn anything new about the investigator's general impressions, observations, and potential risks facing the firm.

For a number of reasons it is essential that the executive management be present during all debriefing sessions. For one, it may be perceived as a lack of interest on behalf of the firm's upper management if they are not involved in these interactive sessions. Additionally, it is essential that the executive management with the authority to allocate resources lead the exit conference and properly demonstrate to the investigator the firm's commitment to managing potential risks to public health identified during the inspection process.

Once the inspection has ended, the firm's RAM strategy continues to play a pivotal role in crisis prevention. If a FD-483 form was issued, the firm must mobilize to develop a corrective action plan (CAP) in response to the deficiencies, unless a GEM Plan was provided, in which case the firms need only reiterate what the plan sets forth and place the investigator's observations within the context of the firm's continuous-improvement plan. This stage is where having initiated a RAM strategy really comes in handy. If a FD-483 was not issued, the firm must make every possible effort not to be lulled into a false sense of confidence.

In addition to the administrative action of receiving an FD-483 at the end of the inspection, there are many other risks that can result from an unfavorable PAI. Table 7 represents some of the negative outcomes that can result if a firm does not implement a RAM strategy prior to the PAI.

Table 7 Potential Risks and Outcomes Associated with the PAI

1. Issue Warning Letters
2. Withhold product approval
3. Suspend NDA, ANDA submissions, and approval
4. Preclude or rescind a government contract
5. Make a nonsuitability determination
6. Initiate or order a mandatory vs. voluntary recall
7. Impose fine(s)
8. Apply the application integrity policy (This is formerly the fraud policy and is invoked when the integrity of data or information filed with FDA is questionable.)
9. Impose criminal and civil penalties
10. Initiate seizure, destruction, or reconditioning of product
11. Impose an injunction decree
12. Debarment
13. Criminal prosecution: felony and other criminal convictions
14. Import detentions
15. Import alerts and suspensions
16. Generate unfavorable publicity and public image

Abbreviations: PAI, pre-approval investigation; NDA, new drug application; ANDA, abbreviated new drug applications; FDA, Food and Drug Administration.

Managing and controlling the actual inspection process begins long before investigators' arrival and lasts well after they have left the firm. Nevertheless, there should be no surprises at the end of the inspection, if during the inspection process the firm's personnel consistently

- hone in on the investigator's concerns by paying close attention to what follow-up documentation is requested;
- convene daily to mitigate risk by evaluating the investigator's comments, responses, questions, and body language during the course of the inspection; and
- stay late, and prepare whatever documentation is needed to support the firm's commitment to continuous improvements and quality.

The key is to assuage the investigators' concerns before they leave the premises and conclude the inspection. Another significant risk intervention tool is the proactive GEM Plan discussed earlier in the chapter.

POST-INSPECTION RISK MANAGEMENT AND CRISIS PREVENTION STRATEGIES

As promised earlier in the chapter, this section will elaborate on the GEM Plan as an important derivative of a comprehensive RAM strategy; and as an initiative that should be launched prior to a PAI, and certainly if the firm's management is aware of prevailing issues that are bound to surface during the inspection. The GEM Plan can and should be used after the firm becomes aware of the specific nature and extent of the citations and the investigator's intention to issue an FD-483. Presenting the plan at the end of the inspection allows the firm to imprint a favorable impression on the investigator's mind and will influence the investigator's development of the Establishment Inspection Report (EIR). This report, along with any documents and photos collected during the inspection will be forwarded to the investigator's supervisor and subsequently to the district compliance officer for further evaluation. It should also be attached to the firm's formal response to the FD-483, which will also be reviewed by the compliance officer.

If a television weatherperson can use computers and Doppler monitors to forecast whether or not it will rain tomorrow, and a hedge-fund manager uses a "black box" to envision the direction of financial markets, and insurance companies can use statistical programs to predict approximately how long an individual will live, doesn't it stand to reason that FDA-regulated industries can also design a "predictive modeling" tool that evaluates a variety of factors in order to determine the regulatory risk facing them? Imagine a proactive, QA program that is able to analyze a plethora of technical data and compliance information, and subsequently forecast the firm's regulatory risks, along with a meaningful compliance rating. Well, there is such a program—it begins with a RAM strategy, evolves into a GEM Plan, and becomes the firm's continuous improvement mechanisms for years to come.

A GEM Plan is a proactive initiative that has the potential to significantly improve an organization's overall compliance and quality status. This initiative is compliance and quality-centered and typically marshaled by the company's executive and quality management. The GEM Plan allows a company to conduct performance analysis, identify its weaknesses, and install exactly what is needed to enhance quality and bring the company into full compliance before the FDA imposes regulatory or judicial action. The GEM Plan objectives and goals must be accurate, far-reaching, and comprehensive in order to convince the agency that the company can launch such an initiative successfully and ensure that it uses the deficiencies revealed during the inspection as a platform for further assessment and improvement. The GEM Plan can be viewed as a strategic initiative for managing crisis before it strikes. During the analysis period, extensive data, practices, information, records, and products are evaluated with the aim of forecasting the firm's regulatory, compliance, and quality vulnerabilities. The comprehensive assessment should cover all critical compliance systems (process validation, personnel training, change management system, etc.) and determine whether or not they meet FDA expectations and current industry standards. Additional warning signals also taken into account and figured in the equation are previous FDA actions, such as the following:

- Does the firm have any outstanding FD-483s?
- Does the FD-483 contain repeat violations?

- Is there an outstanding warning letter?
- Is there evidence of inadequate responses to a previous warning letter or FD-483?
- Is there evidence of commitments to the agency that have not been kept?
- Are there multiple inspections in a short time span?
- Are there multiple inspections of related facilities?
- Is there evidence of nonapproval of recent applications or government contracts?
- Is there evidence that the agency is reluctant to meet with firm officials?

As mentioned in an earlier section, the sum total of this assessment results in the assignation of a risk rating. An essential characteristic of a GEM Plan is the ability to assign a rating that accurately reflects the company's compliance and quality profile, from which a risk rating can be derived. There are a number of possible rating systems that can be employed as part of a GEM Plan.

Another benefit of a GEM Plan is that it can help enlist the district's cooperation during implementation of the plan because the FDA will recognize the firm's proactive, risk assessment, and management efforts. A GEM Plan can integrate compliance issues cited on any FD-483, with continuous improvement activities already underway. This integration will allow the firm to restore its credibility with the agency, the public, and other stakeholders.

Risk Management: The Role of Executive and Quality Management

Another important aspect of the GEM Plan is the role and responsibilities of the firm's executive management. A GEM Plan reflects forward thinking on behalf of the firm's executive and quality management. It demonstrates that they are effectively communicating and interfacing with overall company management, and are committed to a risk management strategy before crisis actually strikes. Involvement from a highly visible and proactive executive and quality management team, with the authority to allocate resources, is the only way to ensure the GEM Plan will succeed.

Mutually agreed upon objectives and goals must be delineated in the GEM Plan. The alliance of and commitment from both divisions (executive management and QA) will invariably create company-wide enthusiasm and motivation. It will also demonstrate to the FDA that there is internal collaboration and executive management commitment to support a long-term, comprehensive, forward quality improvement plan. After all, the FDA shares the same objective—effective risk assessment and management that will ensure high quality, safe, and effective products.

There will be a need to evolve the GEM Plan as it unfolds. Executive management and QA must ensure that the organization remains flexible and realistic. Management's commitment and rapid response to reorganizing, redirecting, and allocating resources, where needed, is the key to success and full compliance. The needs analysis may show that personnel enrichment and skill building are a crucial component to GEM Plan success, in which case, those resources must be committed.

Understanding the Difference Between a Corrective Action Plan and a GEM Plan?

How is the CAP similar to the GEM Plan? Both are exercises in damage control and risk management; however, the GEM Plan is driven by the firm's desire to assess and manage risk internally and proactively.

The CAP is a direct response to the investigator's findings—it is based entirely on what the FD-483 identifies as deficient within the company during the course of the inspection. In contrast, a GEM Plan uses the FD-483 as a launch pad from which to launch a greater initiative—an overall assessment above and beyond what the FDA performed. This assessment has particular appeal to the agency because they know that a typical inspection never exposes all of the firm's deficiencies. The GEM Plan is a compliance-centered and quality-directed initiative totally under the control and leadership of the company's management, not the FDA.

An effective GEM Plan is only as good as the assessment and analysis that preceded it. Curiously, it is easier to enlist the district's cooperation during implementation of a GEM Plan, than during the implementation of an FDA-imposed sanction or CAP. The GEM Plan has the ability to establish a bond of trust between the firm and the local district office. It shows that the firm was proactive and responsible in its assessment of its compliance status. It favorably opens the door for periodic meetings with the local district director and compliance officers to discuss the continuous improvement plan and significant timelines.

Another interesting difference between a CAP and a GEM Plan is that the CAP, designed as a direct response to the FD-483, is, like the investigator's EIR, available through the Freedom of Information Act (FOI). On the other hand, since the firm has, proactively and independent of any inspection, prepared the GEM Plan and presumably contains extensive proprietary information about the firm and its various operations, the firm can choose to seal the document and make it *"confidential and proprietary"* if requested by public entities. It is, however, recommended that the firm prepare a "public information" version of the GEM Plan to accompany the formal response to the FD-483, which will become part of the firm's publicly accessible FOI package. This preparation is an important step in preempting and managing a potential assault to the firm's public image and credibility.

Lastly, whether the company prepares a CAP or a GEM Plan, all risk assessment and management activities must be tracked and reported to the agency on a periodic basis.

POST-APPROVAL RISK ASSESSMENT AND MANAGEMENT PRACTICES

While this chapter is primarily aimed at discussing the various preventive and preemptive measures a firm can take prior to the PAI, it is important to remember that this is not where risk assessment and management measures end. Once on the market, the goal is to remain on the market, and gauge risk at every phase of the product's life cycle. This section will briefly present several risk assessment and management practices that, if properly installed and monitored, can significantly reduce common risks encountered post-commercialization.

One very useful, real-time mechanism for assessing risk, post approval, is the consumer complaint handling system. This critical quality system should be designed to capture, process, and trend quality and medical complaints on a daily basis.

Complaint Handling System

If properly designed, this required quality system could have a great impact on sustaining product quality and forward quality within an organization. A well-designed complaint handling system provides a firm with an opportunity for a 360-degree information feedback loop that can be used to quickly respond to consumer concerns and assess risk in real-time mode. A forward-thinking complaint handling system can also greatly reduce the potential for regulatory and civil liabilities. While the FDA only requires the reporting and trending of certain categories of complaints (serious events that are labeled and not labeled), it would behoove a firm to closely monitor all complaints, whether the FDA requires it or not. Trending consumer feedback and complaints about a product, even when they are not serious, reportable events, could result in learning something new about the way the product performs within the larger population. This information could lead to opportunities for promoting new indications for the product. This has been the case with a number of products that were clinically tested for one indication and discovered (through trending and evaluation of nonserious, nonreportable complaints) to mitigate another disease or condition. There have also been situations where a product was tested and marketed for a particular indication, and later (through the compilation and evaluation of reportable and nonreportable complaints) found to provide benefits for another condition.

Capturing, monitoring and trending complaints during clinical trials, early commercialization, and as long as the product is marketed are FDA requirements; however, in order for complaint handling to serve as a risk assessment and management tool it must adequately harness complaint information and go beyond the regulatory requirements. A mechanism that efficiently harnesses and analyzes complaint information in a timely manner is the only way to truly provide a means for continuous improvements.

Impact of Proactive Quality Assurance Policy on Risk Assessment and Management Strategy

Early in the drug development process, the role and responsibilities of the firm's QA unit must be defined. QA oversight during early development and during all aspects of clinical trial activities is essential. Additionally, QA's close management of activities during the product's initial launch is crucial for successful commercialization, in part, because this is the time that the full extent of the firm's quality infrastructure will be put to the test. All quality systems, from change management and validation, to supplier qualification and annual product reviews, must be evolved to the point of passing a PAI and responding to the many issues that arise during the manufacture and distribution of a commercial product. A RAM strategy must be implemented to efficiently identify and intercept quality and compliance issues before they pose public health risks and assault product marketability. The days of performing one or two internal audits are long gone, and clearly not a useful risk assessment and management approach. Today's regulatory environment requires that QA management assume a *"prevention or risk-based"* perspective that calls for frequent and targeted assessments. Early identification, interception, and prevention are the order in today's fast-paced, FDA-regulated environment. QA must design quality systems and procedures that support prompt detection of problems and ensure the active participation of executive management with the capabilities to allocate resources. There is no doubt that a proactive QA program is necessary for designing a viable RAM strategy and achieving durable compliance.

Risk Management in the Quality Control Laboratory

A very useful quality program, that often gets overlooked, is proactive QA oversight in QC (chemical and micro) laboratories. QA will identify issues related to laboratory management during a routine company audit, or through the identification of repeated OOS results that represent legitimate raw material or product failures. Frequent and targeted assessments of laboratory activities would go a long way in evaluating the overall performance of a given laboratory setting. Issues related to personnel training, data integrity, method inadequacies, instrumentation inefficiencies, journal maintenance, and handling OOS and out-of-trend (OOT) results merit frequent assessments, particularly, in light of the consequences of leaving these critical lab activities unchecked for any length of time. Much of the industry's noted failures and highly publicized hazards can be traced back to deficiencies in the laboratory.

It is essential to develop proactive QA programs that routinely assess the entire spectrum of laboratory activities. The integrity of laboratory data and data authentication should be a high priority. Management review of the analyst's lab journals and worksheets, along with documented evidence that these reviews occur routinely, is also an important QA activity. Periodic performance analysis of a laboratory setting will go a long way in preventing regulatory crisis. Performance analysis of a lab can be achieved by examining a number of areas, along with corresponding data.

Table 8 represents some of the critical activities and data that should be routinely examined to measure the performance of a laboratory.

Periodic quality assessments of these key laboratory activities will ensure that the QC laboratory, frequently thought of as the backbone of the operation, is not compromised, and driving the firm toward a major crisis.

Table 8 Laboratory Risk Management Activities

1. Ensure adequate method validation (accuracy, precision, linearity, range, specificity, ruggedness, robustness, LOD, LOQ)
2. Quality oversight and maintenance of laboratory log books
3. Employ good documentation practices
4. Use appropriate method transfer practices and protocols
5. Ensure consistency with filed commitments (NDA, ANDA, etc.)
6. Confirm instrumentation qualification across the laboratory
7. Ensure instrumentation calibration and PM
8. Use adequate calibration tolerances and standards (NIST)
9. Properly identify pH buffers used
10. Ensure UV and NIR spectrophotometers calibrated for wavelength accuracy
11. HPLC pumps calibrated for flow accuracy, filter change, seal and valve changes
12. Implement SOPs for handling reintegration
13. Ensure HPLC columns undergo performance checks
14. Ensure system suitability established prior to start of analysis
15. Ensure adequate handling of primary reference standards, reagents and working solutions
16. Adequate handling, storage, and tracking of all incoming and outgoing samples
17. Ensure adequate raw material testing (certificate of analysis: specifications, identity testing)
18. Install a retest schedule for raw materials (Do not set arbitrarily.)
19. Ensure adequate system for generation and maintenance of specifications
20. Keep laboratory areas neat and clean
21. Employ sample mix-up prevention system
22. Ensure HVAC and water systems qualified
23. Segregated glassware (clean and verified)
24. Ensure stability testing programs use validated stability-indicating methods
25. Qualify stability chambers
26. Ensure retain sample storage is temperature and humidity controlled
27. Implement system for timely field alerts
28. Ensure SOPs reviewed and updated periodically
29. LIMS qualified and validated
30. Employ adequate laboratory OOS and OOT investigations
31. Employ adequate data security
32. Ensure laboratory change control procedures

(Continued)

Table 8 *(Continued)*

33. Implement data verification and authentication practices
34. Ensure adequate analyst to supervisor ratio
35. Train and retrain personnel, as needed
36. Track number of OOS and OOT results (broken down by category, analyst, and area)
37. Track time between sample log-in and release
38. Laboratory capacity planning

Abbreviations: LOD, limit of detection; LOQ, limit of quantitation; NDA, new drug application; ANDA, abbreviated new drug application; PM, preventive maintenance; NIST, National Institute of Standards Technology; UV, ultraviolet; NIR, near infrared; HPLC, high performance liquid chromatography; SOPs, standard operating procedures; HVAC, heating, ventilation and air conditioning; LIMS, laboratory information management system; OOS, out-of-specification; OOT, out-of-trend.

Risk Management Through Supplier Qualification and Raw Material Control

There are a number of vitally important programs, in addition to the ones mentioned, that once installed, will support long-term, durable compliance and quality, and significantly reduce risk to marketed products.

Since the greatest contributor to product variability is usually presented by its raw materials and components, particularly when provided by an outside supplier, a consistent and thorough supplier/qualification program is vitally important to maintaining product quality and reducing risk to products in commerce.

It is essential to establish a system that allows QC to frequently monitor the variability of all critical raw materials. The ability of QC personnel to quickly detect even small shifts in a raw material's performance will allow QA to anticipate potential product quality problems.

Experience has proven that outsourcing does not always result in cost savings; nevertheless, it's inevitable and unavoidable. It is necessary to partner with suppliers, communicate consistently, and monitor quality levels. Additionally, prior to using any supplier, vendor, or service, a firm or sponsor must adequately qualify the supplier's operation and ensure that it can meet the sponsor's standards and quality requirements. Additionally, it is important to ensure that the sponsor is not inheriting any compliance and regulatory liabilities by virtue of the contractual relationship.

Suppliers and services should be prioritized, in an effort to allocate appropriate resources to those suppliers that pose the largest risk. Contract manufacturers and API suppliers would be high on such a prioritization scheme. Another extremely beneficial quality tool is the quality agreement established between the supplier and the sponsor. This agreement must address and define essential regulatory issues such as the responsibilities of both parties relative to

Table 9 Supplier Agreement Compliance Issues for Contract Manufacture

---

Product release
Establishing specifications
Validation activities (protocol development, actual validation execution)
Personnel interfaces
Retain storage
Documentation revisions
Annual product review compilation
Manufacturing deviation investigations (which group will write the investigation protocol
    and provide final approval)
Product recall
Laboratory investigations (which group will write the OOS investigation plan)
Complaint investigations and follow-up
Notification of nonconformance
Hosting FDA inspections
Agreement updates and revisions
Breach of agreement

Abbreviation: FDA, Food and Drug Administration.

critical compliance activities. Table 9 describes some of the compliance responsibilities that must be addressed in a contract-manufacturing agreement in order to effectively manage risks.

Initial qualification assessments must be thorough in order to realistically measure a supplier's current and future capabilities. Is the supplier's company culture one that the sponsor firm can comfortably embrace and tolerate now and over time? Monitoring or surveillance audits must be targeted and frequent, especially for those suppliers that either pose the largest risk, or have been initially identified as problematic. The FDA expects to see documented evidence that the sponsor has dealt with supplier issues adequately, expeditiously, and thoroughly.

To effectively minimize risk at supplier sites, it is necessary for the sponsor to be proactive with respect to monitoring the changes that occur at that supplier. It is not enough to ask the supplier to periodically update and advise the sponsor of significant changes, as they occur. The sponsor must make frequent and targeted inquiries regarding changes to any aspect of the supplier's operation that could impact the quality of the supply or service. Periodic e-mails or faxes with a limited number of pointed questions might be enough to stay current relative to changes within the supplier's operation. The sponsor must consider and manage all suppliers and vendors as an extension of their company. If the supplier suffers from deficiencies or is the target of a regulatory action, the sponsor will most certainly experience the ripple effect. These days it is necessary to partner with suppliers as much as possible to reduce risk and best manage the contractual relationship.

Manufacturing Deviation Trends and Corrective and Preventative Actions

Needless to say, a manufacturing deviation that results in finished product not meeting its predetermined release criteria should not, from a consumer safety, quality, and regulatory perspective, be introduced into the market place. Fortunately for the consumer and public health, a failed product does not usually find its way into the market. Such a product would be considered adulterated, or contaminated, and nonconforming and might be reprocessed. If reprocessing is not possible, it would have to be destroyed. This, however, might not be the case for a product that did meet its predetermined release criteria, but, during the course of its manufacture, experienced a number of manufacturing deviations. A manufacturing deviation can be defined as

> an incident that occurs during the course of staging, manufacture, packaging, and labeling which is inconsistent with the batch production record, whether or not what has occurred leads to a true product failure.

Imagine that a number of departures from the batch production record have occurred; however, upon performing in-process or finished product testing, all specifications have been met. What might this suggest to quality management? Are there certain manufacturing deviations that, while not causing a true product failure, might indicate potential risk and strongly preclude product distribution? During the development and validation of the process and its methods, there may have been situations that revealed specific manufacturing deviations that were particularly undesirable, while not rendering the material unusable. An example of this might be a finished product that met all in-process and release specifications, but that was stored in bulk drums for a slightly longer period than validated, and under somewhat questionable conditions, but which still meets all specifications upon retesting the bulk prior to finished product packaging. From a regulatory perspective, there is no compliance threat or compliance roadblock preventing release of the product into commerce; however, from a risk management perspective, there may be a legitimate justification for not releasing it, and averting a possible recall.

Another example might be where a specific batch is not in question, rather a number of consecutive batches across a specific time frame. Imagine the situation where a number of batches were manufactured, all met in-process and finished product testing, they have been released to inventory destined for commerce in the months ahead, and, upon a routine batch record audit, it was discovered that all of these batches experienced a myriad of manufacturing deviations. Additionally, it has been noted that there had been some unexpected construction in the facility around the same time frame the batches were manufactured. Environmental monitoring data reveals that there were a number of times during that same period that alert limits were met for several corridors leading to pertinent manufacturing areas.

Once again, while there is no regulatory reason rendering these lots unreleasable, a prudent and risk-based QA professional might legitimately determine that these lots, because of the high incidence of manufacturing deviations and less than optimal environmental conditions during the same period, should not be released into commerce.

A risk-based QA manager might consider evaluating the manner in which manufacturing deviations are investigated. Is there a standardized approach for a manufacturing investigation plan? Is there a uniform report format? Has management offered some guidelines or minimum standards for how investigations should be conducted?

Inconsistent investigational approaches, coupled with a lack of minimum requirements, will invariably lead to some investigations being less comprehensive and effective than others, significantly impacting the quality of the investigations, and leading to inconsistent release judgments and conclusions.

It is recommended that firms, in an effort to reduce the potential for inappropriately releasing product, periodically evaluate manufacturing deviation trends within the context of their unique product lines, circumstances, and environments with the aim of establishing criteria (other than product failure) that would justify not releasing product into market, or withhold until further assessment.

Predetermined criteria, in addition to in-process and finished product specifications that signal QA to possible problems could significantly assist with the interception of potentially problematic product destined for commerce.

The industry has come to rely on in-process and finished product specifications as the only true reflection of a product's integrity, failing to optimize other information that reveals more about the product than whether or not it failed to meet specifications.

Since the cGMP regulations require that the industry perform annual assessments [under 21 CFR 211.180(e)] of their product(s), more frequent assessments of meaningful quality markers such as manufacturing deviations, could effectively support a RAM strategy.

The medical device regulations require multiple assessments throughout the year, of a number of quality markers that are intended to alert executive management of potential or existing quality problems and trends.

While the quality systems regulations require corrective and preventative actions (CAPA), periodic assessments of CAPA trending are not required. Similar to reviewing manufacturing deviation trends, assessing CAPA trends can reveal significant organizational challenges as well as prevailing quality patterns.

When corrective and preventative actions are consistently inaccurate and minimalist, product quality will fall off. Additionally, when corrective and preventative actions are installed, and presumed to be absolute without a short and long-term assessment of the action, many problems can arise. It is important not to assume that the CAPA installed today is necessarily the best practice or solution over the long haul. A long-term review of the CAPA will be the best indicator of its long-term viability.

A periodic evaluation of the long and short-term effects of selected CAPAs, or overall CAPA trends, can reveal valuable information about a company's quality and compliance status and could provide an indication of potential risk for any given period.

Out-of-Specification and Out-of-Trend Data

The annual product review typically specifies the number of OOS results obtained over the course of the year for (in-process and finished product) testing related to any single product. While this is valuable information, a more effective risk assessment and management practice would be to evaluate with more frequency OOS trends related to high volume products.

This information could reveal the QC trends related to a particular product for a given period. Fluctuations in QC activities such as an increase in the number of laboratory-generated OOS results, could signal to QA personnel that potential risks loom. An increase in OOS and OOT data might suggest that there are laboratory deficiencies such as poor managerial oversight, or inadequately qualified laboratory equipment and personnel. A trend of this nature might also point to poorly validated methods. Whatever the genesis underlying troubling OOS and OOT patterns, the information strongly suggests unreliable QC practices that could significantly assault the integrity of the products making their way into the market.

In recent years, the industry and the FDA have evidenced the impact substandard laboratory practices, rampart OOS data, and lack of risk management within QC laboratories has had on marketed products, attempts to commercialize new products, overall company credibility, and compliance status.

Evaluating OOS and OOT results for the sole purpose of annual product reviews does not serve as a real-time risk management measure. A more effective mechanism for intercepting questionable material before it is released into commerce would be to periodically assess OOS trends and OOT results within the context of cGMPs in the laboratory and evaluate their impact on manufactured products.

Periodic evaluations of OOT data is a reliable predictor of oncoming trends relative to raw and in-process material and finished product, laboratory practices, and method or assay migration. Evaluating OOT data shifts toward the upper or lower points in the range and identifying the cause would go a long way in preventing true OOS results and a decline in product quality. Determining the point at which OOT results constitute enough of a risk to product quality that it should be withheld from commerce, until further evaluation, is contingent upon the nature of the product, the manufacturing process, its medical necessity, the speed with which a remediation plan can be implemented, and, ultimately, how risk-averse the firm is.

An effective RAM strategy must establish alert and action limits for the various quality markers, like OOS and OOT results, such that the actions taken— either withholding product or reprocessing it—are timely and part of a continuous improvement effort.

Stability Data and Trends

A failed stability test or trend that points to a fluctuation in the quality profile of a product is an obvious quality indicator that typically results in a field alert, market withdrawal, and product destruction. Trends in stability data are particularly important when dealing with biologic products where, for example, a decline in vaccine potency will invariably continue to decline and become a true OOS. For certain categories of products, the observance of a change in the stability profile, even a minor fluctuation, could suggest more serious problems as the product degrades further. Thoroughly understanding the stability profile of the product can serve as a meaningful quality indicator, not just when an OOS occurs, but when the established profile presents unexplained fluctuations that might suggest withholding the product from the market.

Quality Control Profiles: Raw Materials, In-Process Material (Work-in-Progress, Bulk), and Finished Products

Similar to unexplained fluctuations in stability profiles and OOS trends, changes to raw materials, in-process, and finished product points to any number of problems related to the manufacturing train, process validation, the supplier, or the environment, just to name a few. The greatest variability in any product is presented by its raw materials, particularly when provided by an outside supplier. It is essential to establish a system that allows QC to frequently monitor the variability of all critical raw materials. QC's ability to quickly detect even small shifts in a raw material's performance will allow QA to anticipate product quality over time and prevent potential risks associated with inconsistent material. The Annual Product Review (APR) will capture and report shifts in a raw material's QC profile; however, as suggested above, this is a retrospective review not a prospective, real-time risk assessment. Another excellent tool for preventing product recalls is ensuring that material is adequately qualified before it is introduced as a raw material or substituted for an approved material.

The Supplier Qualification Program

A comprehensive supplier qualification program is another mechanism a firm can use to minimize the possibility that unreliable product will be introduced into commerce. Managing changes within a supplier facility, particularly when the facility is thousands of miles away, can be challenging. There are ways, however, to ensure that change management occurs frequently and effectively. It is most effective when the sponsor assumes the proactive role and becomes responsible for periodically checking up on the supplier and inquiring about changes that might affect the quality of the product. While the supplier agreement and purchase orders typically specify the requirements, along with the caveat that should anything change, the sponsor must be immediately notified, change management should not be left entirely up to the supplier. Changes considered *significant* by the sponsor may not be seen as significant by the supplier.

Firms might consider installing a mechanism that allows the sponsor to periodically, and frequently e-mail or fax a short questionnaire to the supplier regarding changes over a specified period. This mechanism will keep the sponsor apprised of changes that have occurred and permit them to determine what impact, if any, the change(s) may have on the product.

Tracking, trending, and evaluating shifts in specifications and quality of raw materials, in-process or bulk materials, and finished products will provide a failsafe alert system and mechanism with which to monitor a product before it is introduced into the market place, significantly reducing the risk of a firm having to recall product.

With respect to raw materials provided by a third-party supplier, it is essential to monitor, track, and evaluate the performance of the suppliers involved. Qualification of materials, along with extensive qualification of the supplier or vendor is a critical activity for maintaining compliance and quality. Recurring supplier issues can significantly assault product quality over time. Supplier profiles, coupled with adequate monitoring and trending of supplied materials, are useful indicators that can effectively support a RAM strategy.

Specification Drift and Device Feature Creep

For medical devices, there is a phenomenon recognized as *"feature creep"* where the specifications for the design features of a device begin to drift toward a specification limit because of inadequate design reviews, output verification, and change control. This same sort of variability occurs with pharmaceutical and biological products, and without adequate change control, validation maintenance, revalidation, and frequent monitoring of the specifications for any given product can also shift or drift toward a specification limit very quickly.

Product variability can be reduced through constant monitoring, either by statistical process control once enough batches have been manufactured, or through batch-to-batch comparisons of data. Where a specification range has been established, whether for an analytical method or manufacturing process, it is very important to observe fluctuations and drifts in any direction. This fluctuation is an easy and practical indicator that enables QA to anticipate potential product quality issues before they actually arise.

Inadequate change control and regulatory reporting of changes will eventually result in the uncontrolled and unapproved product finding its way into commerce. It is not uncommon for proposed changes to receive incomplete and inadequate assessments prior to the implementation of the change. Additionally, when changes are made to a product's critical parameters such as its specifications, packaging, and labeling without the appropriate notification to the applicable regulatory agency, the product is considered in violation of what was approved. This violation would constitute reason for a recall, cease and desist order, and possible fines from the FDA. Specification drifts and changes in a product's approved parameters can easily occur without adequate change management.

Internal Audit Findings

Similar to complaint trends, internal audit information that reveals reoccurring noncompliance patterns related to product-specific activities, such as manufacturing, process controls, stability testing, and operator performance, should signal QA to potential risks.

Lack of adequate follow-up to product-related audit findings is another clue that there may be a need to hold off on releasing a product until further investigation and remediation occurs.

A ubiquitous failure of many internal audit programs is the lack of a standardized, internal rating system. The results of every audit should receive a rating, and each audit should be measured against the previous audit(s) in an effort to determine whether or not the firm is making the necessary improvements. Evidence of continued digression or lack of any improvement, audit to audit, would serve as a very useful quality indicator and real-time alert mechanism for QA to use as part of its RAM strategy.

Sterility Failures and Trends (Retest Rates)

A sterility failure is sufficient reason not to release product and may eventually lead to the need to destroy it. A periodic assessment of stability test failures and retest trends would alert QA to investigate any possible risks associated with the sterility testing method and the continued distribution of that particular product.

Environmental Monitoring Data

Similar to establishing alert and action limits for the number of OOS and OOT results allowed before the release of a particular product is suspended, establishing similar limits as part of a comprehensive environmental monitoring program can minimize risk. This alert and action data can also form part of a product's release criteria and provide a safeguard against releasing product that may be contaminated or that requires further testing.

Water Data

Water system qualification and re-qualification data, along with periodic water testing will reveal excursions and signal QA to withhold product from the market. Establishing alert and action limits for this crucial raw material will significantly reduce risk to the marketed product.

Returned Goods

Periodic tracking and trending of returned goods, whether or not reentered into inventory, along with a product that required rework or reprocessing is another useful indicator that can serve to alert QA of potential risk to product quality.

Label Reconciliation Practices, Results, and Trends

Adequate label reconciliation is a regulatory requirement, as well as a critical part of QA for all products. Studies have shown that the majority of recalls are attributable to a label mix-up. QA should ensure that as part of the firm's internal audit assessment, a review occurs of its labeling practices, label storage and security, reconciliation, and line clearance. A full understanding of how this activity is controlled should be part of any RAM strategy.

Additionally, if internal audits reveal that there are deficiencies in this area, QA can assess the risk related to product destined for distribution. Recurring label reconciliation problems must be investigated and evaluated against the risk of releasing product.

Formal Line Clearance Practices, Results, and Trends

Similar to label reconciliation, inconsistencies related to formal line clearance procedures can also result in shipping product that may later need to be recalled. It is important for QA to ensure that periodic assessments of this activity occur and that deficiencies uncovered are addressed within the RAM strategy.

Correlation of Interrelated Quality Markers

The benefits derived from comparing information from one quality system, such as manufacturing deviations, with the frequency of consumer complaints or laboratory investigations for that same product can be quite revealing and useful.

Comparisons and a meaningful correlation between compatible quality markers will invariably yield information that can serve to assist QA in its risk-based analysis prior to product distribution.

Annual Product Reviews and Outcomes

As stated above, an APR system is a regulatory requirement and a useful quality tool. Knowing how to develop a comprehensive APR system is just the beginning of having the APR consistently support quality activities. APR information must be used routinely to adequately monitor product performance and play a role in the overall RAM strategy.

Many companies prepare an APR, but do not take full advantage of this tool as a risk assessment and management mechanism. An annual review may not offer the timeliness required to prevent the release and distribution of substandard product; nevertheless, since it is a regulatory requirement, it should be optimally utilized.

Risk Management and the Personnel Component

There is no denying the impact personnel have on product quality. Issues such as training, personnel turnover, staff's core capabilities and knowledge, job expectations, personnel motivation, and leadership all contribute to overall quality and product integrity.

QA must periodically evaluate these factors and determine whether or not they present potential risk to the firm. Frequent performance assessments that examine the ebbs and flows of personnel practices and trends can serve to reveal the potential impact personnel patterns have on product quality and potential risk. Evaluation of personnel activities must be part of a firm's overall RAM strategy.

Concepts in Quality by Design for Drug Development and Manufacture

Ken Morris and Ryan McCann

*Department of Industrial and Physical
Pharmacy, Purdue University,
West Lafayette, Indiana, U.S.A.*

INTRODUCTION

The thrust of this chapter is to describe the process of designing products with the end in mind. Many industries already employ the concepts of quality by design (QbD) in their manufacturing. The pharmaceutical industry has been slow to do this for several reasons that will be discussed in the next two paragraphs. The focus of much of the discussion is on development of new drugs; however, an implementation strategy for generic drugs will be discussed after the concepts are introduced.

Why Are U.S. Drugs So Safe?

To examine the pressures that have delayed the pharmaceutical industry's ability to implement quality, it is instructive to first examine why drugs are so safe today. Drugs in the United States are safe today for four primary reasons. First, we have very strict front-end controls, which are manifest in the validation of the processes by which products are produced. As we shall see, front-end controls are necessary when the processes involved are poorly understood. The second reason is strict change control. This means that any change from the Food and

Drug Administration (FDA)-approved method of production with an unknown potential for damage requires revalidation of the process to ensure quality. The third reason is the compounds that have made up the majority of historical pharmaceutical products have been based on small molecular drugs. These compounds are relatively soluble and relatively stable, at least compared with more modern pharmaceutical compounds like bioproducts, and pose fewer challenges for the formulation, processing, and manufacturing of final pharmaceutical products. Finally, clinical trials were conducted on material made with the approved process. Historically, it was possible to develop robust formulations and products through a largely trial-and-error process. This heavily experiment-based approach to product development was cost-effective when such trial-and-error approaches could be carried out in a reasonable amount of time and at a reasonable cost.

What Is the Problem with Continuing Current Procedures?

The problems result from the static nature of the situation and industry. Front-end controls essentially mean that we have incomplete reconciliation tests of dosage form performance in the clinic. In other words, regardless of the level of control we have, we have no idea what variation in what property results in a significant change, if any, in the clinic. Additionally, this unawareness means that we may be discarding a perfectly good product because it falls outside of a specification that has no clinical significance. This is quite a different situation from looking at the reproducibility of the manufacturing process, even though it is the primary quality metric for pharmaceutical products. Practitioners in the field agree that a large percentage of the product testing and manufacturing controls that are required provide no benefit for either manufacturability or final product quality. However, even if a large percentage of the testing is unnecessary and very expensive, it is impossible to identify which tests should be discontinued without risking product quality, because we don't know which tests are or are not important to the final product quality.

The "don't change anything" culture has arisen as a result of regulators requiring additional tests in response to any change. As a result, the industry has relatively little knowledge of the relationship between change and what constitutes failure. This is a cyclic problem because the regulatory agency's uncertainty in a company's ability to discern important properties has resulted in a very conservative approach toward allowing change to a manufacturing process. This approach, in turn, removes any real incentive for understanding processes for the purpose of identifying key or critical properties and testing those properties to ensure product quality.

Many of these problems may have continued to go unaddressed or even unnoticed had it not been for the increasing complexity of our active pharmaceutical ingredients (APIs). Small, stable, and relatively soluble small molecular organic compounds have given way to larger molecular organics, which typically have low solubility and often significant chemical stability problems. In addition, the variation in crystal and solid forms is on the rise, which makes these molecules more difficult to handle. This variation comes to its height in biotech-derived molecules, which are often proteins or very large molecules with extremely complicated physical chemical properties and often are used in less common dosage forms. Add to this the fact that many compounds being discovered today are extremely potent and will require low-dose formulations and therefore provide additional challenges in uniformity, worker exposure, etc. If current methods are failing now, just imagine how they will fare with these new compounds and dosage forms. Clearly we cannot "cook" our way to success using trial-and-error approaches with these challenges. It requires an engineering and scientifically rigorous approach.

Some Personal Observations

The FDA, in particular the office of pharmaceutical sciences, has led initiatives for the modernization of our discipline. The constraints of budget and resources at pharmaceutical companies combined with the increase in the pressures on the regulatory agencies, in the form of increased numbers of applications from both the branded and generic industries, have necessitated a proactive approach to the challenges confronting those engaged in the development of new drugs. The basic premise reflected in the initiatives is the agency's philosophy that the pharmaceutical industry houses significant technical expertise. The goal is to remove any barriers, real or perceived, in guidances and the review/approval process that might inhibit employing the best science to develop and manufacture pharmaceutical products of the highest quality.

Each of these initiatives focuses on different aspects of product development and will be discussed in this chapter; however, all of them may be described under the umbrella of QbD. It is much like the fable of the six blind men trying to describe an elephant by touch; each "sees" a different part, but it is still all part of the same elephant.

There are many challenges in making such major changes both in the industry and the regulatory agencies. In the end, change comes down to making a common sense and good scientific case for technical issues and relating this case to the business aspect of companies.

A Measure of Variability: Six Sigma

To put into perspective the significance of our lack of understanding of our products and processes to our bottom line, we have only to consider the measure of variability called Six Sigma. As Figure 1 shows, using defects per million units of product as our metric, companies operating at the three to four sigma level may spend as much as 15% to 25% of their revenues to reduce or eliminate defects (1). Pharmaceutical companies have been reported to manufacture at 2.5 sigma, yet release product at 5.5 sigma, according to G. Migliaccio.

- Defects Per Million
  - ➤ One Sigma 697,672.15
  - ➤ Two Sigma 308,770.21
  - ➤ Three Sigma 66,810.63
  - ➤ Four Sigma 6,209.70
  - ➤ Five Sigma 232.67
  - ➤ Six Sigma 3.40
- "Companies operating at a level of *3-4* sigma may spend as much as 15-25% of their revenues to reduce or eliminate defects." (GE, basics_of_six_sigma.pdf)
- Pharma companies manufacture at 2.5 sigma and release at 5.5 sigma! (G. Migliaccio, Pfizer)
  - ➤ Discarding good material
  - ➤ Shotgun testing

Figure 1 Six Sigma and the Pharmaceutical Industry.

What this means is that if we cannot manufacture to a higher standard (i.e., less defects), we must be discarding good material and at the same time spending inordinate amounts of money on shotgun testing without really understanding the value of the testing, as discussed earlier.

FDA INITIATIVES

Much of the logic required to address our challenges was captured in the FDA document "Pharmaceutical cGMPs for the 21st Century: A Risk-Based Approach" (2). As an industry that seems to follow the lead of our regulatory agency, it was logical to begin with a program to develop a science- and risk-based approach to product quality regulation, which incorporates an integrated quality system approach. The five main subdivisions of that document are shown in Figure 2, with the first two items devoted to QbD, i.e., a risk-based orientation and science-based policies and standards. This chapter focuses primarily on

1. *Risk-based orientation*
2. *Science-based policies and standards*
3. *Integrated quality systems orientation*
4. *International cooperation*
5. *Strong public health protection*

Figure 2 Five main subdivisions of "Pharmaceutical cGMPs for the 21st Century: A Risk-Based Approach."

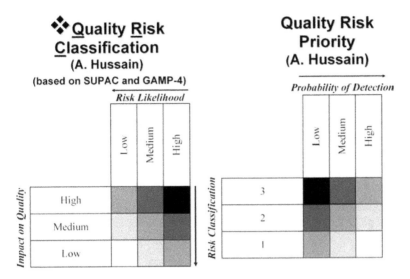

Figure 3 Quality risk classification and priority tool.

those two points, but clearly the remaining points follow from the first two and are incorporated as appropriate.

Focusing for a moment on risks, Figure 3 shows a quality risk classification and priority tool developed by Ajaz Hussain, the former deputy director of the Office of Pharmaceutical Sciences at the FDA. The risk classification of a "type" of failure is not flexible if based on data; e.g., if the compound can hydrolyze to an inactive form, you can't change the molecule at this stage but, with the probability of detection high due to monitoring and modeling of the process, the risk likelihood and priority can virtually always be improved, e.g., by knowing that the compound can hydrolyze at a predictable (modeled) rate and the real-time moisture (monitoring).

Science-based regulations mean less "no value-added" testing and regulatory revision. Additionally it acknowledges how *statistically ludicrous* it is to use one bio batch, three validation batches, and 10 tablets for content uniformity (CU) and/or dissolution. Currently, we use the same strategies for testing and insuring quality for low-low, high-high, or anything in between (Fig. 3)! This is clearly not the most efficient way of ensuring quality or maximizing the efficiency of our processes. The idea is to assess the risk and match the value-added testing to the level of risk.

Formulation/Process Design Issues

Having discussed the need for changes in product development and manufacturing, it is important to point out that many of the skills and skill sets required for QbD already exist in our scientific community. All good formulation scientists

The Goals are the same: e.g., Solid Oral DF

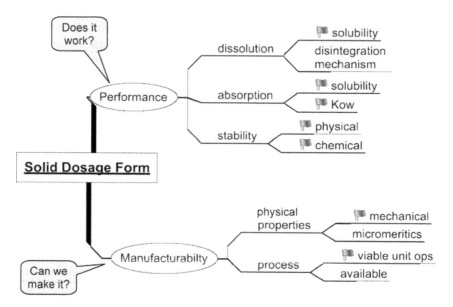

Figure 4 Formulation and process development activities leading to the manufacturing of solid oral dosage forms.

and engineers have always known that a formula is not independent of the process that will ultimately be used to turn the formula into a product. Even during API characterization, developing a formula implies an expected process or at least a range of processes. For example, you would not use a compaction aid for a lyophile or a disintegrant in a solution ready for injection product.

In Figure 4, we see that there are basically two goals from formulation and process development activities leading to manufacturing. The first is to make sure that the dosage form works by reliably delivering the API where it is needed at the rate needed. The second goal, although it precedes the actual use of the product, is to ensure that the product can be manufactured reliably and that it performs as required to meet the first goal. This is obviously tied to the clinical studies that must precede the introduction of a new drug. However, clinical testing ends eventually, and other metrics must be applied to maintain the consistency of the product.

Communication and Design

The difference about the new direction that good manufacturing practices (GMPs) are taking is that they use historic knowledge (this can include personal, literature knowledge, information from vendors, etc.), as well as knowledge to be developed to design products that perform the way they are required to for the patient. The knowledge that needs to be gained will be discussed in later sections of the chapter, but it includes modeling at all levels of sophistication, as well as heuristic knowledge. Equally as challenging as understanding our processes is the task of making sure that everyone in the development and manufacturing train gets the knowledge they need when they need it. This task includes the forward as well as backward transfer of knowledge and requires generating relevant data, turning that data into knowledge, and using the knowledge to make decisions.

ICH Initiatives

There is help from the International Committee on Harmonization (ICH). Although most of the initiatives discussed in this chapter are those coming out of the FDA, clearly the international community is just as aware of the challenges facing drug developers. It is obvious that in a global economy there is no practical way that modernization in just one region of the world will accomplish these goals, as companies must basically abide by the least advanced or most stringent requirements rather than develop different development strategies for each region of the world.

The ICH consists of regulatory agencies and the research-based pharmaceutical industry in the European Union, United States, and Japan. The World Health Organization, the European Free Trade Association (which is represented by *Swissmedic,* a regulatory agency in Switzerland), and *Health Canada* also participate in the ICH as observers.

ICH activity is administered in the spirit of agency/industry cooperation, and the ICH process is science based as well as predictable and transparent. The ICH consists of a steering committee and expert working groups. The steering committee and the expert working groups meet twice a year. The steering committee is the committee in which administrative and policy issues are discussed, while in the expert working groups, technical guidelines are discussed and prepared.

The common technical document (CTD) is really at the heart of the ICH initiatives, and its structure represents the combined thinking of all of the parties involved (3). The committee recognized from the very start that a uniform filing format was needed for harmonization. Figure 5 shows the basic structure of the CTD, sections of which will be discussed to illustrate how the CTD facilitates quality by design activities.

From the perspective of nonclinical development and chemistry manufacturing control (CMC) issues, the quality overall summary (QOS) module 2.3 is really the heart of the CTD. This section includes the rationale for the product design and process design, the underlying scientific principles employed in the design, and summarized critical data supporting the rationale. Pointers from the QOS to more complete data representations in module three are necessary to facilitate the review process. A good QOS should allow the reviewer to fully understand the logic behind the decisions made in the design of the product as well

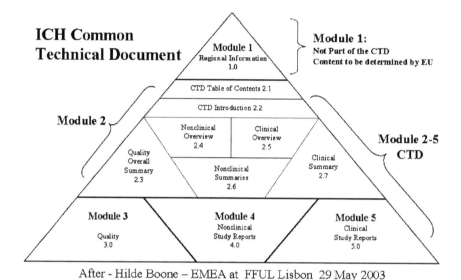

After - Hilde Boone – EMEA at FFUL Lisbon 29 May 2003

Figure 5 Basic structure of the common technical document. Adapted from Ref. 3.

as the level of process understanding by the sponsor. QbD is all about using the understanding of the product in the process in order to control the process to ensure product quality. While the CTD format is not highly prescriptive; ICH guidelines (Q, S, and E) have been generated to supplement the CTD in the areas of quality, safety, and efficacy to aid sponsors in understanding the goals of the CTD.

ICH Q8

For QbD, ICH Q8 is the most immediately relevant guideline ICH Q8: Pharmaceutical Development, ICH Step 4—Note For Guidance On Pharmaceutical Development-(emea/chmp/167068/2004) (4). Some guiding principles from the ICH Q8 guideline are as follows:

> "The information and knowledge gained from pharmaceutical development studies provide scientific understanding to support the establishing of specifications and manufacturing controls. Changes in formulation and manufacturing processes during development should be looked upon as opportunities to gain additional knowledge and further support establishment of the design space."

It is clear that the intent of the new initiatives is to foster the scientific method as our primary process driver for pharmaceutical development. This includes taking advantage of the natural or applied variation in materials and methods to explore as fully as possible the impact of these variations on product quality. The guideline goes on to say, ''The manufacturing process development program should identify the critical process parameters that should be monitored or controlled (e.g., granulation endpoint) to ensure that the product is of the desired quality.'' The principles of process analytical technology, which will be discussed, are clearly foreshadowed and are required to be able to execute the intent of ICH Q8 and QbD. Once a process and product are understood, critical variables identified, and strategies for monitoring in place, the guideline states the spirit of how product specifications should be determined: ''Product specifications are based on mechanistic understanding of how formulation and process factors impact product performance.''

Process Analytical Technology

The process analytical technology (PAT) guidance from the FDA (5) has been the subject of much discussion and perhaps some misunderstanding. It was the first major guidance in the QbD thrust since the PAT guidance is an enabler for QbD. It is crucial to understand that PAT is not just about process analyzers. The primary goals of the PAT initiative are to encourage and facilitate three things: process understanding, process monitoring, and process control. Until one understands their process, there is no way of identifying what variables are critical in determining whether the process is performing properly. Appropriate monitors for a process cannot be identified until the process critical variables have been determined. Monitoring these relevant variables must be done in a timely manner. Having said that, the method of monitoring is completely at the discretion of the scientists and engineers, and the method should be based on the science for the variable being monitored.

The key to effective monitoring is to collect information at short enough intervals compared with the entire processing time so that adjustments can be made to the process in order to maintain control (the third goal of PAT). For example, for fermentation processes that may take many days, a temperature, CO_2, or pH measurement every 12 hours may be sufficient to control the process. It is important to understand that using a sophisticated sensor without controlling the process is not PAT and will not make the sponsor eligible for a PAT filing with the FDA.

Implicit in the PAT concepts is that some sort of quantitative model is employed in the control strategy. These models may be based on detailed first principles models for processes understood at a very fundamental level, all the way to models based on prior or heuristic knowledge. The list shown in Figure 6 gives a range of types of models that may be employed. Irrespective of the model selected, it must have the characteristics of being quantitative, and include the assumptions implicit in using the model and the associated range of applicability. An example of this will be discussed later.

Modeling to Support Design Work Flow

There is a hierarchy of modeling that recognizes the evolution of models from early inception to actual application on a scale process. Starting with the design

PAT Requires Modeling

- Physical-Chemical
    - Thermodynamic (phase transformation)
    - Kinetic (degradation)
- Engineering
    - Heat transfer (drying)
    - Mass transfer (drying)
    - Momentum transfer (blending, bed dynamics)
- Statistical
    - Chemometric (all multivariate processes)
    - Prior knowledge (all)
- Phenomenological
    - Empirical
    - Semi-empirical

Figure 6 Types of models used for PAT applications.

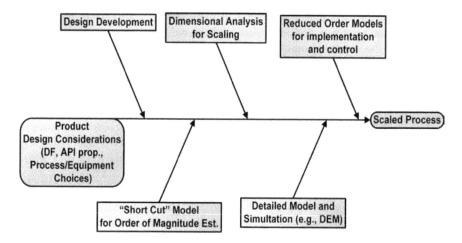

Figure 7 The evolution of models from early inception to actual application on a scaled process.

considerations (Fig. 7), a shortcut model may be developed to capture the order of magnitude behavior of a process. Typically, at this stage, dimensional analysis will be performed to identify dimensionless groups of variables that are scale independent and could be used for the scaling process. A detailed model and simulation is ideally the next step, where all of the variables and assumptions of the model are included. Finally, a reduced order model is constructed by identifying those variables in the detailed model that are found to have an impact on the final product. It is the reduced order model that is then used for the control of the final scaled process. The reduced order model is still dynamic and captures the essence for implementation of control and will also continue to evolve as more data are collected on the process and understanding increases. This model embodies the concept of continuous improvement, which is critical to maintaining the quality of the product over its lifetime. Continuous improvement will be driven by the variation in inputs that will necessarily occur and through capturing of learnings from outside the process, e.g., post-marketing clinical information that may require alteration or adjustment of the model.

CURRENT VS. QBD DESIRED STATE

In many respects, these new initiatives are really freeing the scientists and engineers in our companies to practice their discipline as they have been trained. Let us contrast how the current CMC review process works versus how it would be for a sponsor using the concepts of QbD to attain the so-called desired state.

From Figure 8, it may be inferred that the real goal is to have the same tools that scientists and development engineers used to develop a product and process also be employed by the reviewers as they assess the level to which the sponsor has supported their case.

How does the CMC review process have to change?
Current vs. *QbD Desired State*

- Companies may or may not have info but it's not always in the filing

- Reviewers must go through cycle of info requests and questions

- Companies may or may not have clear scientific rationales for choices but are not always sharing it. (no incentive!)

- Reviewers must often "piece together" data and observations to discover the rationale for a spec, method, formula, process, etc. (=Frankenstein)

- Reviewers are analyzing the data they often must tease out of the company

- Companies include needed data with filing and could share it prior to the filing

- Companies include the data analysis to produce meaningful summaries and scientific rationales

- Reviewers assess the rationales and summarized data presentations as satisfactory or not

Figure 8 Current vs. Quality by Design Desired state for CMC reviews.

Development in the Desired State

If we now couch the total product development time line in terms of QbD, we begin to see a process that more closely approximates the thought pattern which good development scientists and engineers typically employ. There are three major divisions during the development life cycle, viz., *early development*, where recommendations for the dosage form and liabilities are assessed; the *preliminary design space phase*, where the initial process and formulation design are developed; and finally the *transfer* of the final design space to manufacturing. Although this process may not seem like a significant departure from the current paradigm, the underlying assumption is that, unlike current practices, the shift in paradigm requires that all aspects of the development pipeline be represented in each of the stages. The relative contribution of downstream activities will be less at the beginning and increase as time goes on, while the opposite would be true for early development activities; however, all interests need to be represented at the appropriate level throughout the entire process. This ensures that design decisions made early on are realistic for what is occurring downstream and that downstream activities can be fed back to aid in future design decisions.

If we consider a typical set of activities that must be performed during development as shown in Figure 9, we realize that while the specific activities may not change, there can be no silos that isolate early- from mid- or late-stage development.

Early Development: The Preformulation/Materials Science Time Period

Let us examine some of the assumptions and implications of the foregoing treatment using small molecule solid oral dosage forms as our example for the early development stage where preformulation (PF) and materials science engineering (MSE) activities are performed.

We virtually always know what our desired dosage form type is before we start PF (if it is to be administered orally, intravenously, etc., and usually if it is a direct compression or lyophile, etc., specifically desired). The PF/MSE efforts must be focused on attributes of the API that *potentially* impinge on the final selection of processes and therefore formulation. Once known, these attributes dictate the *possible* process selection and/or facilitate the use of possible but nonoptimal processes (e.g., balancing kinetics if wet granulation must be used on a moisture/heat labile compound).

At the end of PF/MSE, assembly of the preliminary development plan should therefore be possible. This plan includes the best choice(s) of process, preliminary formulation recommendations, and the associated risks. This is the preliminary list of API attributes pending final identification of critical dosage form performance attributes and dose ranging. The logic behind such a list is a result of the innate variability in raw materials, which includes the API, as well as excipients that will be used. If raw materials never varied in their

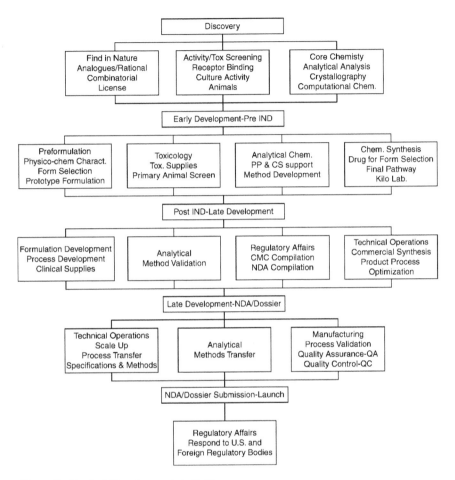

Figure 9 Typical Development Activities.

properties, then as long as the formulation and process were carefully selected and held constant, there would never be failures. Because raw materials do vary, it is clear that one cannot hold a process constant with variable input and expect the product to be constant. In fact, we transfer the variability directly from the raw material to the product if we hold all else constant in the process. So, with knowledge of the desired dosage form and the relevant properties of the API in hand, one may begin to select a process and ultimately the formulation that will both perform well in manufacturing and dosage form performance.

A valuable approach for both risk assessment and for implementing the design concepts discussed is inductive risk assessment. Tools such as the Ishikawa diagram, an example of which is shown in Figure 10 for a granulation process, allow the development scientists and engineers to systematize potential risks given the possible processing pathways available. Such diagrams and tools

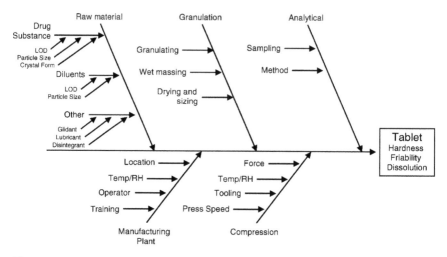

Figure 10   Ishikawa diagram for granulation process.

are recommended in ICH Q9, which is a guidance devoted to risk assessment and management, an integral part of QbD.

To ensure that information generated to make decisions in the early time periods is communicated downstream, appropriate informatics and information management must be incorporated from the very beginning of the process. This can be viewed as the first stage with all subsequent stages being supported by a framework or bed of informatics and information management (Fig. 11). The distinction between information management and informatics is that information management refers to the actual transfer of information, while informatics includes data analysis and transforming the information for use in making decisions. Again, it is assumed that all disciplines are involved at every stage to the degree necessary, and the information management facilitates these activities.

Digging deeper and still using our small molecules solid oral dosage form example, we can further divide the early activities, as shown in Figure 12. Here, we see there are biopharmaceutics, physical chemistry, mechanical and micromeritic, and chemical dimensions to our activities. It is the combination of the data collected during this period and the models used to predict behavior on the basis of mixing rules and prior knowledge that lead us to the preliminary development plan. This preliminary development plan should be generated rapidly and not delay progress; even at this early development stage a preliminary plan of this type is not only possible but necessary.

What Are the Three Big Biopharmaceutics Preformulation Questions?

For the biopharmaceutics dimension, the three questions generally regarded as most critical deal with how well a compound is absorbed, how toxic the compound is, and what a likely dose would be. Each of these questions has tools for addressing the

What is the <u>*Desired State*</u> and What Should Happen in the PF/MSE Time Frame?

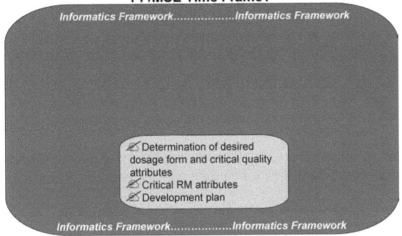

Note: these are time stages only, ALL disciplines must now be involved at EVERY stage of decision making!!!

Figure 11 The PF/MSE Time Frame in the Desired State.

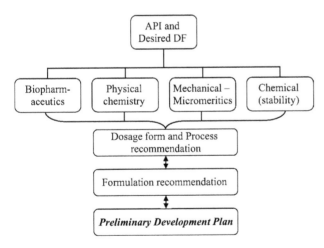

Figure 12 A further division of early development activities leading to the Preliminary Development Plan.

early estimation of these properties. For example, the Biopharmaceutical Classification System (BCS) and the Modified Absorption Potential (MAP) methods (6,7) are widely used to assess potential compound absorption. Toxicity is assessed in animal models, while more recently data-driven quantitative structure-activity

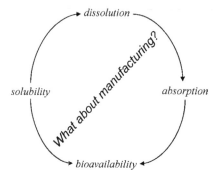

Figure 13 Solubility is a critical property for estimation of absorption and bioavailability, but it is also important when selecting the processing route.

relationship (QSAR) methods are being developed for the efforts of the FDA (8). Likely doses may often be estimated from knowledge of other compounds in the same class; however, this will ultimately be determined by dose ranging studies. The top three physical chemistry questions center on the solubility and solubility improvement of the compound in solvents and pH's of interest, the chemical stability of the compound alone as well as in mixtures of excipients, and the physical stability of the compound once a solid form is known and the thermodynamics is understood.

Impact of Solubility

As shown in Figure 13, solubility is typically considered a critical property for the estimation of absorption and ultimately bioavailability. However, solubility may also be a critical factor to be considered in the selection of a processing route, particularly if a compound is moisture and/or thermally labile. This is obvious for chemically unstable compounds; however, the solubility may also be important in facilitating or mediating phase transformations between solid forms and is a good example of the type of understanding that is consistent with the QbD concepts and with the expectation of regulatory agencies.

The mechanical and micromeritic questions for solid oral dosage forms have to do with powder flow, moisture sensitivity, and compatibility. Powder flow will be a function of particle size distribution, particle shape, and how easily the powder will charge under mechanical stress. These characteristics must be considered when designing a dosage form and selecting a process. Modeling of powder flow has received more attention in the past 10 years employing advanced techniques such as Discrete Element Methods (DEM) to predict angle of repose and agglomeration (9). Moisture is always a potential problem for solid oral dosage forms, depending upon the hygroscopicity and the hydration state of the solid regardless of it being the API or excipient. Compaction and mechanical properties in general are perhaps the least explored in the early stages of development. This can lead to many avoidable problems downstream and may be addressed by a combination of existing techniques from materials science and the pharmaceutical literature, such as Hiestand indices.

Setting Specifications

The philosophy in setting specifications is captured in part in ICH Q6A. The concept is simple in that it assumes the accurate identification of the critical attributes and then applies the logic of the ranges of the attributes that will result in acceptable manufacturing and dosage form performance. This requires "initial" or "interim" specifications based on the logic provided above and any preliminary formulation data, but these must be revisited in light of subsequent development work. In this formalism, the early raw material data must be fed forward to facilitate the formulation/process development studies and backward to facilitate the setting of meaningful raw material acceptance/release specifications, thus avoiding unneeded specs. A clear implication of this logic is that attributes not yet identified as critical must be included to the extent possible in the design of experiments (DOE) activities as development of the design space proceeds. If sufficient variable material is unavailable, the variable in question should be "left open" in the design so that subsequent manufacturing data can capture the impact, and the results may be fed backwards to modify specifications. In the end, the process will be controlled to a desired endpoint; however, the range of raw material attributes may dictate if the desired endpoint is possible, given the above analyses.

Raw Material Concepts to Employ and Use in Development and Assessment

Of course, it is not always easy to know what is important or to be able to measure it, but if the scientists are following the logic, they should be able to more rapidly identify problems at the very least. Many times, indirect evidence will be sufficient to determine the impact and/or track a raw material property. A possible approach is suggested below to illustrate the spirit of the activity.

- *Estimate* on the basis of the science (data, prior knowledge, models, etc.) which properties of the raw materials (API and excipients) may be important for the manufacture and/or performance of the desired dosage form.
- *Assess* on the basis of the science which properties may be impacted by a selected processing train and the potential impact on manufacture and/or dosage form performance.
- *Identify* whether or not an analytical method exists to follow properties of interest, keeping in mind that
  - not all properties can be easily measured, and
  - not all properties that are measurable are important.
- *Design* (or look for) experiments to test the criticality of the properties, including as much variability as feasible.
- *Execute* the experiments.
- *Interpret* the data to establish the potential impact.

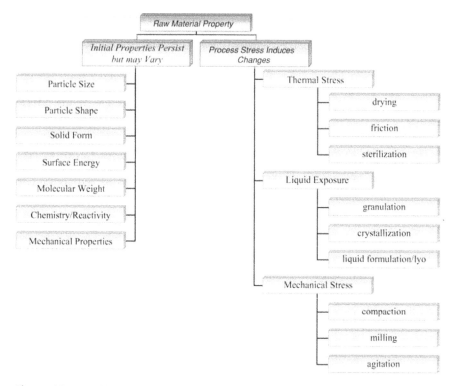

Figure 14 Classification system for the impact of raw materials on manufacture and dosage form performance based on if the incoming raw material property is likely to change during processing.

More specifically, we could classify the impact of raw materials on manufacture and dosage form performance by API versus excipient, or by the unit operations that were going to be used. However, a simple system based on whether the incoming property of the raw material was preserved through the process was selected (Fig. 14). It is not complete or magic, but it has some advantages for illustration of the concepts. The flow chart facilitates the exercise of assessing whether a property is likely to change during processing or behave according to its initial state. It relies upon general categories of processing stresses, and it may be applied to any unit operation in light of the liability of a given type of change.

Some examples of raw material properties that may initially and can potentially impact performance/manufacture are given in Figure 15. This is by no means a complete list, but it is only to illustrate what is meant by "Initial Properties that may persist through processing."

Processing-induced changes in raw material properties refer to unintentional or poorly understood changes occurring during manufacture (Fig. 16). Unintentional means that you would like to retain the properties of the starting

Initial Properties Persist :
Causes - Effects

- Particle size
- Particle shape
  - Wettability
- Mechanical properties
- Solid Form
- Pre-treatment of
  natural products

- Dissolution
- Mechanical strength
- Shelf life

Figure 15 Raw material properties that can potentially impact performance/manufacture of final dosage form.

Processing Induced Changes:
Causes - Effects

- Phase transformation
- Chemical degradation
- Amorphization
- Particle size change
- surface modification
  (coating, roughness,
  etc..)
- Sorption to surfaces
- Tribo-charging
- Changes in MW

- Dissolution
- Concretion
- Mechanical strength
- Potency
- Impurities
- Immunogenicity
- Shelf life

Figure 16 Process induced changes to raw material properties that can impact the final dosage form.

material but something happens, e.g., a phase change. Poorly understood refers to the common case when something changes during processing and produces a desired effect but you do not know it is occurring (e.g., particle size reduction). As the incoming raw material properties vary, the process may not produce the desired results. Without understanding of the process, no one knows why the desired result is not met (i.e., no assignable cause for deviations).

Critical Attributes for All Formulations

Having identified the types of properties and how they might be affected during processing, it remains to determine which may be critical attributes or properties for manufacture and/or dosage form performance. Some of these decisions are easy since stability, solubility, absorption, and assay will be critical for virtually all dosage forms. Beyond those attributes, performance of different processing trains will rely on different properties of the raw materials and the raw materials' response to the different stresses placed on them. Returning to our solid oral dosage form example and further restricting it to tablets, we may illustrate a logical thought process that may help to narrow the list of potentially critical attributes and help decide where to expend resources early in development.

Critical Attributes for Wet Granulation

For example, if an API must be processed by wet granulation, the properties of solubility, solution stability (physical and chemical), wettability, and excipient interaction are all potentially critical. These may well be interrelated as a chemically labile compound's solubility combined with the exposure to granulating fluid will determine the degree of degradation it undergoes. Similarly, solution-mediated transformation from one solid form to another will be proportional to those factors, as may the degree of interaction with excipients. Whether the API degrades on association with the excipient may be a separate component of the process.

Critical Attributes for Direct Compression or Roller Compaction

If the liabilities of wet granulation outweigh the potential benefits, a "dry" process must be selected. The most typical alternatives are roller compaction (i.e., dry granulation) or direct compression. Each potential process now presents its own set of potential critical properties/attributes. For roller compaction, the flow characteristics may not be as critical as they will be for direct compression, but deformation, bonding, and their ultimate impact on wetting and dissolution must be assessed. Flowability may be the deciding criterion for selection of one dry technique over the other. The level of first principals understanding of flow is improving but still immature. However, useful relationships to assess flow include the Carr indices, which provide the ability to "rank" the relative flowability of powders. Powder flow impacts blending, flow from hoppers, and die filling. Therefore, insufficient flow may well rule out a direct compression process. For either direct compression or roller compaction, the compactability of the API may be critical to designing both the dosage form and selecting the process. So, as with all of these properties, they must be ruled in or out as potentially important. Our lack of first principles understanding of molecular organic solids and granular systems means that we again rely on qualitatively understood relative rankings of the mechanical properties. The Hiestand indices are a tool developed around materials sciences methodology to anticipate bonding as well as the potential for failure on ejection or impact (10). Having determined the critical characteristics of the API and selected the most logical process, the excipients may be selected on the basis of their own functionalities and the levels at which they are required. This selection, of course, depends on the final dose that may still be in flux at this stage, but at least the major characteristics have been identified and understood as thoroughly as possible.

From Where Does All This Material Come?

For new drugs, some of the analyses described above are not routinely done because of tradition or most often because of lack of material early in development.

The testing that takes the most material is the mechanical properties and physical form screening. There are mechanisms to obtain sufficient material for these physical tests. Specifically, the material provided at the multigram level to formulate toxicology supplies can be used for these nondestructive tests prior to making the solutions/suspensions required for the toxicology studies. The FDA's only interest is the results of the toxicology studies, so as long as care is taken not to introduce foreign substances into the material, there is no barrier to its use. Also, during the development of a commercial synthetic pathway, lots may fail to meet acceptance criteria such as purity, yield, or particle size distribution. While these drugs would certainly not be used for toxicology or clinical studies, they may be perfectly acceptable, as is or modified, for mechanical testing and/or form screening. The data must, of course, be viewed as preliminary. However, extrapolation of the mechanical properties of materials at various densities (solid fractions) to a fully dense material is a useful baseline for comparing material produced later and should be a relatively invariant parameter.

What About Generics?

The generic industry operates under different time constraints than the branded industry. This is because at the time of abbreviated new drug applications (ANDA) filing, the generic company will not have had time to execute the procedural elements discussed. However, a medium to large generic company may have dozens of similarly developed dosage forms and should be able to use the Bayesian concept of "prior knowledge" as the basis for the development of a preliminary development plan and rationale. Using this approach, the API characteristics, which may be largely known from the literature or determined quickly during this stage, can be reconciled with similar materials and formulations/processes used in the past. For example, an API with similar characteristics to previous APIs may logically be matched with a similar excipient mix and processing train. Keep in mind that, although the API may be known, the formulation may have to be developed from scratch for a generic product, so there is really no realistic way to implement QbD for generic manufacturers without including risk assessment, both inductive and deductive.

Mid- to Late-Stage Development: Formulation and Process Development Time Period

In the mid to late stages of development, the primary focus will be on identifying the preliminary design space (Fig. 17). Note from Figure 17 that there should be open communication between these time periods, and although they are separated for the purposes of description, they should really represent a continuum punctuated by the Investigational New Drug (IND) for new drugs or the Abbreviated New Drug Applications (ANDA) for generic drugs. Also note that the same requirements for information management and informatics apply to all stages of development to be discussed.

What Should Happen After PF/MSE?

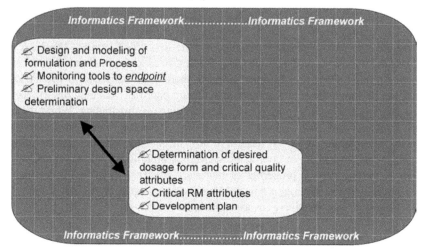

Note: these are time stages only, ALL disciplines must now be involved at EVERY stage of decision making!!!

Figure 17 Activities during the Formulation and Process Development time period remain connected to earlier activities.

Starting with the best choices from the preliminary formulation recommendations in the preliminary development rationale, as well as their associated risks, the real task is to define the range of operating parameters and raw material properties that, when combined in manufacturing, reliably produce high-quality product. This is done by taking knowledge of the dose and critical quality attributes discussed, the design of a practical process train, and the understanding of the process variables (i.e., unit operations models), and exploring the maximum variability feasible for both formulation components (particularly the API) and the process. During this period there will, of course, be refinement of both the formulation and the process train selected. The design space should be as portable as possible, but this issue will be discussed later. The idea is that the more fundamental the understanding one has about the process and the formulation, the easier the scaling of that process will be. During this time, tools for monitoring the process should be constantly explored and exploited. It is the marriage of the identification of critical quality attributes with the selection of an appropriate monitor that is critical for what will be the process control strategy.

PAT Is the Enabler for the Desired State

Clearly, PAT is an enabler for QbD. It enables the use of scientific understanding as the basis for decision making, as opposed to blind regulation or convention, and requires the use of controls to rational endpoints rather than being arbitrarily controlled by time. PAT empowers the use of appropriate sensors and identified process critical control points (PCCPs) to generate statistically significant high-quality data and rules for real-time release while providing incentives for innovation and encouraging continuous improvement. In short, you cannot really have a design space without PAT.

However, not all monitoring needs to be done online or by sophisticated spectroscopic sensors in order for PAT to be useful in process control. For example, inline temperature measurement suffices if the temperature is a PCCP. A sophisticated monitoring system that does not allow control may be useful in troubleshooting and refining processes. However, this type of system does not constitute control and by itself would not warrant review under the PAT guidance. Analytical method substitution is allowed under current conditions with the appropriate method reconciliation and validation. PAT, as an initiative, was designed to provide the incentive for innovation by removing barriers to change and insisting upon science-based control and continuous improvement.

Monitoring of Critical Attributes

The use of sensors to monitor critical attributes as a way to control endpoints determined by design space models is desirable. Again, qualifying for PAT status implies that process understanding is used to control the processes, not just more accurately document failures. This requires that critical attributes be monitored to allow the process to be modified on a near real-time basis. The PCCPs will, in part, be the manifestations of the raw material variability, and the monitors used will have to be either directly or indirectly sensitive to the raw material critical attributes. An example table matching attributes to logical analytical methods or "eyeballs" is provided in Figure 18 for solids. The table is not meant to be exhaustive. The criterion for an appropriate technique is that the science of detection must be consistent with the information desired. It must be understood that the analytical signature of raw materials may be complex and only understood in terms of multivariate analysis. The table in Figure 18 also shows the sensors associated with unit operations common to a variety of processes.

ICH Q8—Pharmaceutical Development

This scientific understanding establishes the design space as formally presented in ICH Q8. In these situations, opportunities exist to develop a more flexible regulatory approach to facilitate

- risk-based regulatory decisions (reviews and inspections);
- manufacturing process improvements, within the approved design space described in the dossier, without further regulatory review; and
- "real-time" quality control, leading to a reduction of end product release testing.

ASTM E55.O2 RM WG: Monitoring of CAs

Unit Operation	What is monitored	Best Eyeball(s)
crystallization	conc, size, shape, form	IR, Raman, XRD, Laser Diff.
liquid/semi solid mixing	homogeneity, particulates	UV, Laser Diff

blending
    gravity
    high shear
	homogeneity	theifing, NIR, LIF, others

granulation
  high shear
  fluid bed
  roller compaction
  spray drying
	torque, power, moisture, size, density
	temp., moisture, size
	ribbon content uniformity, hardness
	size, form
		equip. output, NIR, thermal, sonic
		NIR
		equip. output, NIR
		lasentec, NIR

drying
  fluid bed
  other
	temp., moisture
	temp., moisture
		NIR, thermocouple, RH
		NIR, thermocouple, RH

particle size reduction
  micronization
  milling
	size, temp, form
	size, form
		Laser dif, lasentec, NIR, Image analysis

sizing		Sieve, NIR, Laser, Image analysis
compaction	hopper uniformity, tablet CU, hardness	NIR
coating	weight, thickness, dissolution	NIR
capsule filling	CU, weight	NIR, auto weigh
sterile fill	particulates, microbes	RMD, conductivity, laser diff
lyophilization	Tg, moisture, temp., vac.	"smart" freeze dryer, conductivity

Figure 18 Analytical methods for monitoring critical attributes in various unit operations.

Q8 goes on to define the design space:

"Design Space: the design space is the established range of process parameters that has been demonstrated to provide assurance of quality . . . .
. . . Movement out of the design space is considered to be a change and would normally initiate a regulatory post approval change process (4)."

ICH Q8 also provides direction on what is considered a change as we move from procrustean specifications to the design space (i.e., there's no free lunch).

Many industries use, and have used for decades, design space concepts. It is a combination of our justifiable caution and our inertia that has prevented us in the pharmaceutical industry from using these same concepts. Using a design space means that varying elements of the process and raw materials is permitted as long sufficient understanding of the process and products have been demonstrated. Also, we have the ability to detect the endpoints so that the final product is assured to have the high quality demanded. If you are operating within the design space (including processing options, sites, materials, etc.), you are producing quality product. If you are outside the design space, it is a deviation and/or out of compliance (producing "adulterated product" in regulatory language).

However widely used design spaces are in other industries, it is instructive to see an example of design spaces from one of our commonly used unit operations in order to gain a level of familiarity and comfort with the concepts.

Example: Fluid Bed Drying

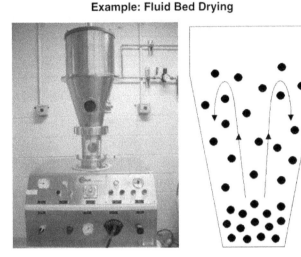

Figure 19 Fluid bed drying equipment and internal schematic diagram.

- Let us take a basic example of fluid bed drying of a granulation
- Assume that you can vary (*the hair dryer model*)
    - ➤ the inlet air temperature
    - ➤ the inlet air flow – i.e., volume per unit time
- Assume you can assess the moisture content in real-time (NIR, RH.
    temp, etc…) and *control* it to the desired endpoint
- How do you construct a design space?
    - ➤ Empirically
    - ➤ First principles
    - ➤ (combinations)

Figure 20 Fluid bed drying model assumptions and potential critical variables to be monitored.

The example we have selected is that of fluid bed drying (Fig. 19 shows a typical fluid bed dryer and a schematic of its internal working). It was chosen for two reasons. First is that fluid bed drying is one of the most widely used unit operations in solid oral dosage form processing, and the other is that we may use this to illustrate the use of both empirical and first principles design spaces. In fact, it is one of the few unit operations for which the first principles are relatively well understood, but we will discuss this as we proceed.

As introduced earlier and as with any model, one must clearly state the assumptions and identify as clearly as possible the potential critical variables to be monitored. Then the question arises, how do you construct the design space? As alluded to in Figure 20, design spaces may be constructed empirically, from first principles, or from a combination of the types of models presented earlier.

Experiments are conducted at various values of the
temperature and volume (DOE) to produce a plot that
shows the range that gives the desired moisture level

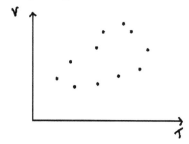

Figure 21 A plot of inlet air temperatures and air volumes which produce the desired
moisture content.

Empirical Design Space

Starting with the construction of an empirical design space, we must first
delineate our assumptions and incorporate our understanding of what may be
critical into the process. In this case, the following assumptions are made:

- You can vary the inlet air temperature.
- You can vary the inlet air flow, i.e., volume per unit time.
- Moisture content is a critical attribute (and response variable).
- Moisture content can be assessed in real-time [using whatever technique
  is appropriate or available, e.g., near infrared spectroscopy (NIR), RH,
  temp, etc.].
- You can control the moisture content to the desired endpoint as determined
  by the response variable or surrogate monitoring value.

With these assumptions in hand and considering the limitations of the
equipment being employed, a DOE approach may be used to probe and populate
the design space as shown in Figure 21. For each point, the processes are run until
the desired moisture level is achieved as measured by the monitor selected. The
values for each experiment are then plotted and the plot is analyzed to identify
the boundaries of the design space, as shown in the hypothetical plot (Fig. 21).

The boundaries may be defined by a fitted function or by a conservative
regression (where the lower bounds of the perimeter values are used for each
segment to find the boundary), as shown in Figure 22. Alternatively, simply
connecting the dots may be sufficient (as long as the interval is small enough)
with a margin of error dictated by the reproducibility of a single experiment.

Depending on how the boundaries were defined, inequalities may be
generated that provide a quantitative relationship for representing the design
space. In any case, it is the inscribed area that represents the empirically derived
design space as shown in Figure 23.

The connected "dots" form straight lines, each
with its own equation; V = m T + b

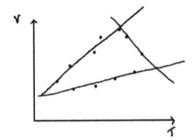

Figure 22 Defining the boundaries of the empirical design space.

By making inequalities, the design space is found
V ≤ mT + b or V ≥ mT + b depending on the line

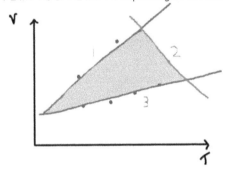

Figure 23 The shaded area represents the empirical design space which is defined by inequalities which give the quantitative relationship.

While empirical design spaces do facilitate endpoint-based control, compared with first principles design spaces, they require more data to confirm boundaries, make it more difficult to assess out of specification (or out of space) results, and, although they may aid in scale up, they still leave us with a higher level of uncertainty. It should also be noted that the use of DOE and the understanding of the process range variability that produces acceptable material is not a new concept. The beauty of the design space concept is that it leverages much of the expertise already in our companies and discipline.

First Principles Design Space

A first principles design space relies on a more fundamental understanding of the process. For drying, this refers to the heat and mass transfer characteristics of the process, as has long been studied by chemical engineers (Fig. 24). There are two primary stages to drying; the first stage, or constant rate period, describes the loss of

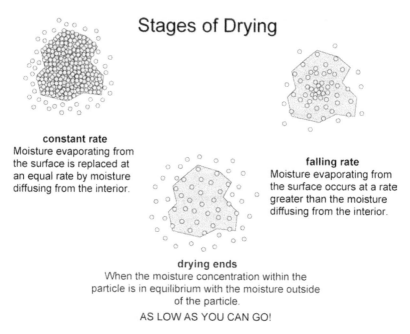

Stages of Drying

constant rate
Moisture evaporating from the surface is replaced at an equal rate by moisture diffusing from the interior.

falling rate
Moisture evaporating from the surface occurs at a rate greater than the moisture diffusing from the interior.

drying ends
When the moisture concentration within the particle is in equilibrium with the moisture outside of the particle.
AS LOW AS YOU CAN GO!

Figure 24   Schematic of heat and mass transfer stages of drying.

excess or surface moisture, while the second stage describes the process of moisture diffusing from inside the granule to the surface before evaporating. The third stage refers to the limit of drying one may achieve depending upon the balance between the relative humidity of the incoming air and the conditions in the dryer.

As shown in Figure 25, the physics and mathematics of these processes have been worked out for many years (11). The first stage of drying is where we are losing only surface moisture as a linear relationship with time and all of the energy being supplied by the heated incoming air is used to evaporate this loosely associated moisture. This stage is referred to as the heat transfer limited stage. The relationship in the second stage between moisture loss and time is exponential. This stage reflects the fact that energy must be used in order to defuse the moisture to the surface before it evaporates and then separates from the surface. This stage is therefore referred to as the mass transfer limited stage. As the combined graph in Figure 25 shows, the moisture loss follows this linear and then exponential relationship. Meanwhile, the temperature is constant during the heat transfer/evaporative stage because all the energy is being used to evaporate moisture and then increases as energy is being absorbed by the solid to facilitate the diffusion of the moisture to the surface.

Variables in Fluid Bed Drying Models.   Implicit variables, in the relationships just described, are listed in Figure 26 along with the assumptions used for this study and applied to fluid bed drying in general. Now we begin to see the strength of the first principles method and that the relationships we have

1st Principles Based Design Space E.g.

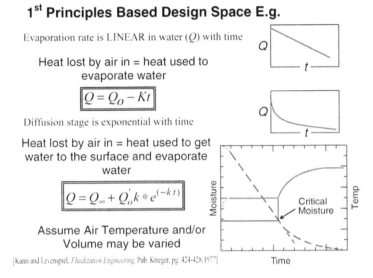

Evaporation rate is LINEAR in water (Q) with time

Heat lost by air in = heat used to evaporate water

$$Q = Q_O - Kt$$

Diffusion stage is exponential with time

Heat lost by air in = heat used to get water to the surface and evaporate water

$$Q = Q_\infty + Q_o'k * e^{(-kt)}$$

Assume Air Temperature and/or Volume may be varied

[Kunn and Levenspiel, *Fluidization Engineering*, Pub. Krieger, pg. 424-428, 1977]

Figure 25 The Physics and Mathematics of drying.

Variables in Fluid Bed Drying Models

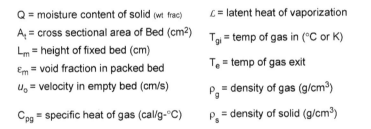

Q = moisture content of solid (wt frac)

A_t = cross sectional area of Bed (cm^2)

L_m = height of fixed bed (cm)

ε_m = void fraction in packed bed

u_o = velocity in empty bed (cm/s)

C_{pg} = specific heat of gas (cal/g-°C)

$\mathcal{L}$ = latent heat of vaporization

T_{gi} = temp of gas in (°C or K)

T_e = temp of gas exit

ρ_g = density of gas (g/cm^3)

ρ_s = density of solid (g/cm^3)

- It is also generally safe to assume that
  - the temperature gradient within a particle is small
  - the temperature is constant throughout the bed
  - the vapor gradient within a particle is small

Figure 26 Assumptions and variables used in the first principles model of drying.

employed allow us to change the predictions on the basis of differences in our equipment, material, and in how we use the material in the equipment, e.g., the height of the fixed bed.

In our example, we will be using NIR as the monitoring strategy (Fig. 27). NIR is extremely sensitive to water absorption and therefore provides an ideal monitor for moisture. However, the other monitors of temperature and relative humidity may be equally as effective.

Using the NIR calibrated for moisture content, we can compare the actual data collected with the predicted values from the equations solved using the

Uniglatt -MM55 NIR at Purdue

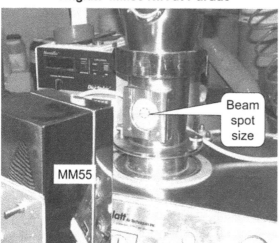

Figure 27 Near infrared experimental set-up for fluid bed drying.

Fit to Evaporative Drying Model for APAP Formulation

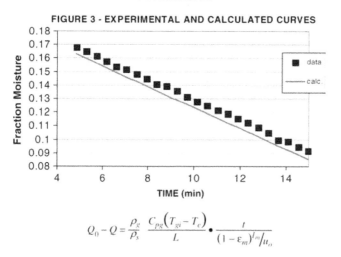

$$Q_0 - Q = \frac{\rho_g}{\rho_s} \frac{C_{pg}(T_{gi} - T_c)}{L} \bullet \frac{t}{(1 - \varepsilon_m)^{l_m}/u_o}$$

Figure 28 Experimental moisture data compared to the calculated values from the heat transfer equation.

variables specific to our system. As we see from Figure 28, the agreement is excellent for the first stage of drying and a similar relationship is found for the second stage of drying. It is this type of predictability upon which we rely to first validate our models and then to assess what might cause a deviation from validated models with changes in materials, scale, etc. This is not to say that the

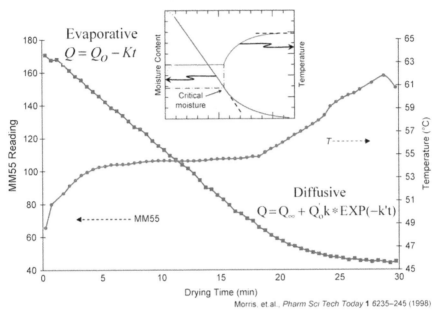

Figure 29 Real-time moisture data via NIR and outlet air temperature for an acetaminophen and starch granulation.

models themselves may not have to be modified in the face of more complete data and/or training sets; however, we do gain the advantage of knowing what we expect to see so that we know, at the very least, where to start with our scaling activities.

In Figure 29, we see the near real-time moisture data via NIR and the real-time outlet air temperature data for an acetaminophen and starch granulation being dried (12). The inset shows the qualitative predictive behavior, and we observed that the real data reflect this almost exactly. Figure 29 illustrates a point often confused during discussions of PAT. The endpoints we use often consist of two components, i.e., as the system reached the endpoint or plateaued with respect to the parameter being monitored and the value at that endpoint. It may be sufficient for some unit operations only to establish that the operation is finished in order to release material for the next unit operation, while for others it may be necessary to assess the actual value. Depending upon the criticality of the particular operation being modeled, it is often necessary to know both the stage of completion and the value for use in a design space.

The next four figures illustrate how using first principles relationships can facilitate both the construction of the design space and its use in scaling activities (13). Figure 30 shows a response surface that corresponds to the moisture content as a function of time and inlet air temperature. A slice parallel to the moisture and

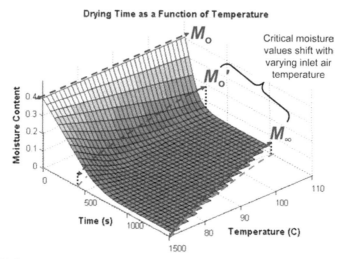

Wildfong, et.al, in prep

Figure 30 Response surface for moisture content as a function of time and inlet air temperature.

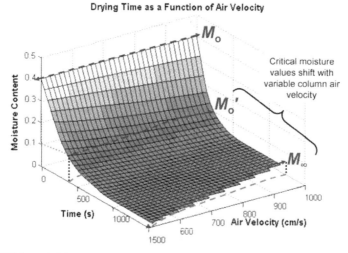

Wildfong, et.al, in prep

Figure 31 Response surface for moisture content as a function of time and inlet air velocity.

time plane at any temperature generates a drying curve qualitatively described earlier, although it varies quantitatively as predicted by the relationships.

The next figure shows the same type of plot; however, this time for the inlet airflow and the same qualitative and quantitative behavior is predicted (Fig. 31).

The third figure shows the drying time necessary to attain the desired moisture content plotted as a function of the two adjustable variables, inlet air temperature and velocity (Fig. 32).

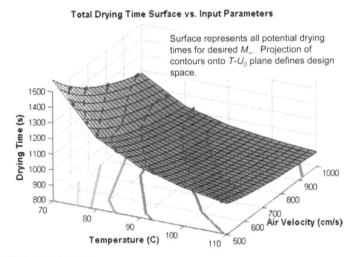

Wildfong, et.al, in prep

Figure 32 Response surface for drying time to reach a certain moisture content as a function of inlet air velocity and temperature.

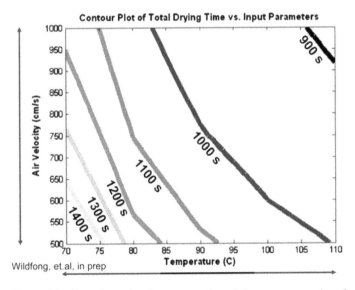

Wildfong, et.al, in prep

Figure 33 Two-dimensional representation of the response surface for drying time as a function of inlet air velocity and temperature.

Note that the response surface represents the desired final moisture content and level curves projected from that surface onto the temperature—air velocity plane can be used as a two-dimensional representation of the three-dimensional situation. We can use this as a visual representation of our design space (Fig. 33).

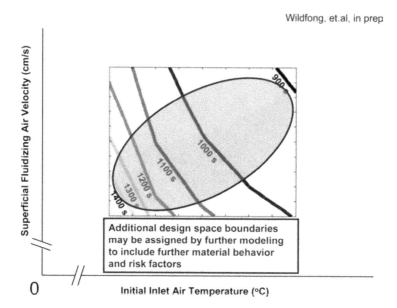

Wildfong, et.al, in prep

Figure 34 Additional design spaces can be placed on top of the original design space to produce a production space.

Depending on the limitations of the equipment being used, the relationships will be solved over different ranges of possible values for each of the variables, as shown in Figure 34. Also shown is the concept of combining design spaces. For example, if granule size is a critical attribute in addition to moisture level, as it often is, the combination of variables resulting in the required moisture level and granule particle size distribution must be combined to produce what is sometimes called the production space.

Scheduling and economic decisions may also be facilitated using a first principles approach. In the Wildfong example, Figure 35 shows the time savings that would be realized as a function of combination of variables selected. This approach may be useful in everything from determining the number of shifts required in manufacturing to the amount of energy used in the processing.

Out of Space(s) (No Free Lunch)

As discussed earlier, there is no free lunch, so data that fall outside of the design space have to be treated according to the logic of the science underlying. If they are outside of the space, the limits of the design space may have to be revisited. We must evaluate the experimental error, as always, and if the error bars still put you within the design space, then you are producing material within the area you have demonstrated produces quality product. If not, then either your model is off because you have an incomplete model (lack of understanding of unidentified critical attributes), your system has changed so that it doesn't fit your model (this

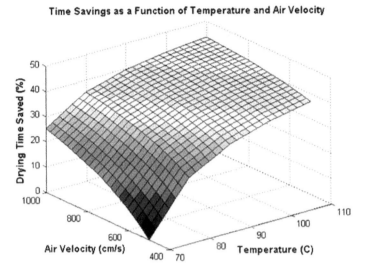

Wildfong, et.al, in prep

Figure 35 Response surface for time savings as a function of inlet air velocity and temperature.

means that your process or materials have changed), or your monitoring measurement is off. In any case, troubleshooting such deviations is facilitated by using the models. Whatever the level of sophistication of your models, troubleshooting must be done in order to identify the root causes. Another potential advantage is that models may highlight variables often considered critical which, in the light of sufficient data, turn out to be noncritical and therefore need not be monitored on a permanent basis.

Do First Principle Models Really Work?

Figure 36 above shows an application of first principles modeling and implicitly the use of a design space (14). It shows two NIR drying curves for ibuprofen and starch granulations. The upper curve shows a drying process conducted at 60°C inlet air temperature with a fixed airflow. This is a typical process for ibuprofen, as the melting point of racemic ibuprofen is approximately 75°C. Taking advantage of that part of cooling, as predicted in the heat transfer limited stage of drying, one may dry at a higher inlet air temperature, in this case 80°C, to hasten the drying and then reduce the temperature at the onset of mass transfer limited drying to avoid compromising the product. The lower curve shows this process and results in a 50% decrease in the drying time.

We know that design spaces can work, but we are limited by our level of understanding of some of our unit operations. Having said that, barriers of expertise may not be as significant as many people fear. The tools and activities for

Fast Drying Trials of an Ibuprofen Granulation

Comparison of Average MM55B Values
Between Fast Drying and Traditional Drying

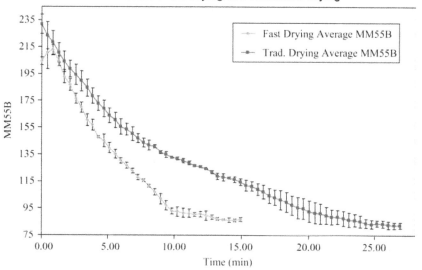

Figure 36 Fast drying application of first principles model for fluid bed drying.

understanding the range of variability of the process that produces acceptable material, and therefore design space, is not a new concept. However, little effort is typically expended in design space development, because historically there have been no regulatory incentives, and compressed timelines seemed to have made exploring the variability impractical. While it is true that constructing design spaces may require more upfront effort, it is not necessarily true that this requires a significant increase in the amount of time. In some ways, it is a matter of convenience to plead lack of time when there is always time to fix a problem after it occurs.

Combined Risk Assessment

For the benefit of the development process and to aid in the filing process, a combined risk assessment at the end of the development is essential. Shown in Figure 37 is the mate to the Ishikawa diagram introduced earlier; Robert Menson (personal communication) has called it an "influence matrix," which I believe to be the most accurate description. This table summarizes all of the unit operations and their potential impact on quality attributes identified during the development process. In the QOS, such a summarized assessment may be linked to specific examples both in module two and module three of the CTD to facilitate the reviewing process and construction of the QOS.

Influence Matrix

Unit Operation Quality Attributes	Blending	Granulation	Drying	Blending (lubrication)	Tableting
Dissolution		Not sensitive -exp't		Control ranges	Not sensitive
Disintegration					
Hardness		Not sensitive – exp't			Design space
Friability		Density controlled			Design space
Assay					
Content Uniformity	NIR Control				
Appearance					
Identification					
Water		RH control	Moisture monitoring/ control		
Microbiology					
Degradation	Not sensitive – prior knowledge				
Stability					

Figure 37 A process influence matrix summarizes all of the unit operations and their potential impact on quality attributes.

Manufacturing: Scale-Up and Manufacturing Time Period

The transition from development to manufacturing will include many of the same specific activities that are currently in place, i.e., formulas and processes must be transferred and batches at scale must be produced. However, we really are transferring the design space with all the understanding that has hopefully proceeded (Fig. 38). In addition, representation from manufacturing during development should serve to minimize some of the historical redevelopment that has plagued our industry. Clearly, there will be differences with our initial scale design space. Knowledge gained in manufacturing batches should help to produce larger and more representative training sets of data for use in refining and parameterizing our models. This knowledge leads to a major change in mentality where the three-batch "custom" is replaced by generating the number of batches needed in the judgment of the manufacturing engineers, likely in a highly truncated DOE, to provide a sufficient level of confidence for release to the public. Specifically, starting with the knowledge of the formulation, process, the design space, and monitoring strategies should make preparation for producing batches at scale much cleaner than they often are now. Once the agreement and/ or variation from the developmental design space is assessed and adjustments are made to produce the initial scale design space, release between unit operations or even final release is made on the basis of the endpoint determination of process critical control variables as allowed by the design space. In addition, traditional

Then What Happens?

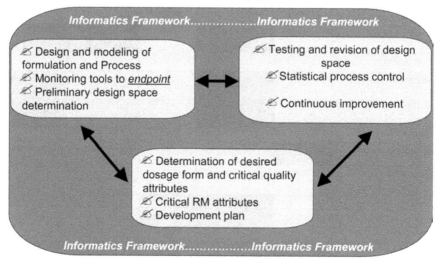

Note: these are time stages only, ALL disciplines must now be involved at EVERY stage of decision making!!!

Figure 38 During the Scale-Up and Manufacturing time-period the design space is transferred and continued communication with previous activities occurs.

tools of statistical process control, process capability, and trending must still be employed. When incorporated into an overall control strategy, including feedback loops, these methods can play a key role in identifying trends and making prospective corrections for subsequent manufacturing, as well as enriching data sets used for continuous improvement of multivariate statistical tools used in many of the monitoring strategies employed.

Fluid Bed Drying Example: The Scaled-Up Process

Returning to our drying example, using the same first principles-based design space monitoring strategy, a scaled study was executed. As Figure 39 shows, the error in our estimate on the small scale is in the neighborhood of 12%, and on the large 100-kg scale the system behaved as predicted.

Scale-up in the engineering science typically requires the construction of new equipment that preserves the key characteristics and dimensionless quantities determined at the small-scale. This is typically not the case in the pharmaceutical industry, because of both custom and the impracticality of designing new equipment for every product. However, if the material-critical properties may be monitored and the endpoint is associated with the material reaching demonstrated values, much of the equipment dependence may be minimized.

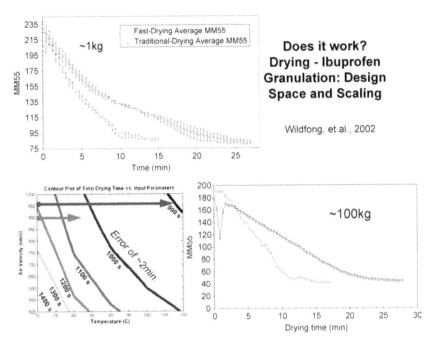

Figure 39 The fast drying method was applied to a production scale granulation successfully.

SUMMARY

It is through QbD that the desired state of development may be achieved. This desired state can be summarized in the following seven points:

1. Product and processes are *designed* from the most fundamental principles available, to produce a product with desired critical attributes.
2. Relevant data is generated and non–value added testing are eliminated.
3. Data are captured and turned into information and communicated up/down stream for timely decision making.
4. Sources of variability are understood and incorporated into a multilevel design space.
5. Processes are controlled by monitoring PCCP and adjusting in near real-time to a desired endpoint.
6. Improvement is continuous, employing appropriate elements of statistical process control, process capability, stability, etc.
7. Open communication means that real-time release and decreased regulatory hurdles are then possible.

While achieving only the last point would represent a considerable success, there are still significant technical barriers in three main areas (Fig. 40).

Quality by Design Inhibitors: gaps in our knowledge and ability to implement QbD concepts in a comprehensive manner

1. Material Science

- Small molecular organics and macro-molecules are not well understood

2. Process Modeling

- First principles engineering models are lacking

3. Information Management and Informatics

- The ability to quickly turn data into information for decision making and process control is a challenge

Figure 40 Technical barriers to the integration of Quality by Design into the pharmaceutical industry.

Small molecular organic and macromolecules used in the pharmaceutical industry are very different from the traditional materials studied by material science engineers (metals, ceramics, etc.) that are well understood. The difficulties in understanding small organics or macromolecules come from their relatively weak inter/intramolecular interactions when compared with the strong ionic and covalent interactions in metals. Also, these organic molecules have increased degrees of conformational freedom, which makes the modeling of their behavior more difficult. Finally, these types of materials have less well-defined computational tools (force fields, mechanical properties, etc.).

First principle engineering models of pharmaceutical unit operations are needed to predict process scale-up, but we are currently lacking the quantitative predictive models to do so. Without these models, the impact and level of process related stresses cannot be anticipated even if the material science is known for the material. As discussed earlier, development using QbD concepts is dependant upon understanding how process stresses will affect raw material attributes so that the proper processing trains can be selected.

Information management and informatics is also a major hurdle for implementation. For QbD to work properly, we must be able to get information from numerous sources and formats to "talk" to each other so that decisions can be made and process control applied. Also, first principle engineering models need to be included to make sense of the data from the unit operation design spaces. The information, both gathered and produced, must be fed both forwards and backwards along the processing train to maintain control. The tools to do all of these tasks must be developed before we can proceed with QbD implementation.

To further complicate the issue, all three barriers must be overcome together since each barrier requires knowledge and information from the others. However, there is good news. We now know what the problems are and we already do most of what is required to implement QbD. All we need now is some

The challenges are significant but are now within our grasp:

Organizational
  - Silo busting, coordination, incentives for improvement
  - Internal regulatory modernization/reconciliation
  - Informatics

Regulatory
  - FDA initiatives, training
  - Global lack of harmonization

Scientific
  - Materials science
  - Unit operations modeling
  - Informatics

Figure 41 Overview of how the remaining organizational and scientific challenges will be addressed.

basic science/engineering knowledge and for some people to get out of the way so that change can occur.

Many of the remaining organizational and scientific challenges will be addressed by industry and academia. There is current research in all of the scientific areas and progress is being made toward removing these topics as challenges. The regulatory challenges will be met through FDA initiatives like the PAT guidance and cGMPs for the 21st Century. The lack of global harmonization should be improved through the efforts of the ICH and their guidances (Fig. 41).

Overall, there are many challenges that must be overcome before the desired state will be met, but progress is being made. The ideas and methods of QbD are already in place within the industry, and it is just a matter of time until the desired state becomes the current state of pharmaceutical development.

ACKNOWLEDGMENTS

We wish to thank the Consortium for the Advancement of Manufacturing of Pharmaceuticals (CAMP) and the Indiana 21st Century Fund.

REFERENCES

1. GE Medical Systems Healthcare Solutions. The Basics of Six Sigma. Available at: http://www.gehealthcare.com/twzh/prod_sol/hcare/pdf/basics_of_six_sigma.pdf. Retrieved July 30, 2007.
2. FDA. Pharmaceutical CGMPs for the 21st Century—A Risk-Based Approach: Final Report. Available at: http://www.fda.gov/cder/gmp/gmp2004/CGMP%20report%20final04.pdf. Retrieved July 30, 2007.

3. FDA, CDER, CBER. Guidance for Industry M4: Organization of the CTD. ICH, August, 2001. Available at: http://www.fda.gov/cder/guidance/4539O.pdf. Retrieved July 30, 2007.
4. ICH. ICH Harmonised Tripartite Guidance: Pharmaceutical Development Q8: Step 4 version. Available at: http://www.ich.org/LOB/media/MEDIA1707.pdf. Retrieved July 30, 2007.
5. FDA, CDER, CVM, ORA. Guidance for Industry: PAT—A framework for Innovative Pharmaceutical Development, Manufacturing, and Quality Assurance. Available at: http://www.fda.gov/cder/guidance/6419fnl.pdf. Retrieved July 30, 2007.
6. Lobenberg R, Amidon GL. Modern bioavailability, bioequivalence and biopharmaceutics classification system. New scientific approaches to international regulatory standards. Eur J Pharm Biopharm 2000; 50(1):3–12.
7. Sanghvi T, Ni N, Yalkowsky SH. A simple modified absorption potential. Pharm Res 2001; 18(12):1794–1796.
8. Kruhlak NL, Contrera JF, Benz RD, et al. Progress in QSAR toxicity screening of pharmaceutical impurities and other FDA regulated products. Adv Drug Deliv Rev 2007; 59(1):43–55.
9. Pottmann M, Ogunnaike BA, Adetayo AA, et al. Model-based control of a granulation system. Powder Technol 2000; 108(2–3):192–201.
10. Hiestand EN. Rationale for and the measurement of tableting indices. In: Alderborn G, Nyström C, eds. Drugs and the Pharmaceutical Sciences: Pharmaceutical Powder Compaction Technology. Vol 71. New York, NY: Marcel Dekker, 1996:219–244.
11. Kunii D, Levenspiel O. Fluidization Engineering. New York, NY: Krieger, 1977.
12. Morris KR, Nail SL, Peck GE, et al. Advances in pharmaceutical materials and processing. Pharm Sci Tech Today 1998; 1(6):235–245.
13. Unpublished data, with permission; Peter Wildfong, Robert Cogdill, Duquesne University.
14. Wildfong PLD, Samy A-S, Corfa J, et al. Accelerated fluid bed drying using NIR monitoring and phenomenological modeling: method assessment and formulation suitability. J Pharm Sci 2002; 91(3):631–639.

Equipment Cleaning During Pharmaceutical Product Development and Its Importance to Pre-Approval Inspection

Lisa Ray

Eli Lilly and Company, Indianapolis, Indiana, U.S.A.

INTRODUCTION

The pharmaceutical product development process is long in duration and complex in nature. This complex process is simplified and shown in Figure 1 using terms that will be referenced throughout this chapter.

This chapter is specifically focused on equipment cleaning and its importance during the many stages of pharmaceutical product development from early formulation development through pre-approval inspection (PAI). Specifically, this chapter outlines general equipment cleaning requirements and translates those requirements into acceptable equipment cleaning strategies for pharmaceutical products throughout the various stages of product development. Particular focus throughout the chapter will be applied to the basic contents of a pharmaceutical manufacturer's cleaning master plan (CMP). A sufficient CMP, when written and followed, will provide documented evidence that appropriate controls are in place to prevent cross-contamination, lending support to a successful PAI.

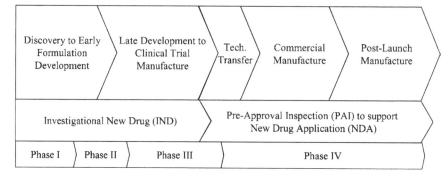

Figure 1 Pharmaceutical Product Development Matrix.

The federal government publishes the Code of Federal Regulations (CFR) as a codification of general and permanent rules. Title 21 of the CFR is reserved for rules governing the Food and Drug Administration (FDA). CFR Part 211—current Good Manufacturing Practice (cGMP) for Finished Pharmaceuticals Subpart D—Equipment Sec. 211.67 provides the requirements for equipment cleaning and maintenance (1,2). In accordance with 21 CFR 211.67, the International Conference on Harmonization (Q7A) has issued recommendations on equipment maintenance and cleaning (section 5.20–5.26) for compliance and safety that include very similar requirements with more elaboration on specific details (3). Again, there is a reiteration that written procedures shall be established, and that detailed cleaning agent selection and preparation, responsibilities, schedules, and cleaning acceptance limits are documented with rationale.

The FDA has these cleaning requirements/standards in place in order to ensure the safety, identity, strength, quality, and purity of the manufactured pharmaceuticals and to prevent cross contamination between different manufactured lots. This cross-contamination concern applies not only to the active pharmaceutical ingredient, but also to residues left from cleaning solvents and detergents. Other potential sources of contamination include, but are not limited to, the following: active ingredient, degradation products, excipients, organic solvents, products of equipment wear, environmental dust, residual rinse water, particulates, and lubricants.

At various stages throughout the pharmaceutical product development process, the FDA expects a controlled and validated equipment cleaning program, justification for cleaning acceptance limits, ongoing monitoring of the equipment cleaning program to ensure effectiveness, and good training for all activities (including equipment). Early phase clinical studies can be conducted before the details of the product formulation and/or product process have been fully defined and validated. Some of the good manufacturing practices (GMPs) are not as clearly applicable to the manufacture of clinical trial material for phase I studies. For reference, phase I studies are typically open label studies that impact approximately 50 healthy volunteers where the study objective is primarily safety. Phase I clinical trials typically include the initial introduction of an investigational new drug (IND) in humans. This trial is in contrast to phase II and phase III trials, which may involve substantially greater numbers of subjects being exposed to the drug product. Phase II and III studies aim at testing the effectiveness of the drug product. Although the cGMPs apply to the preparation of any drug product administered to humans, the FDA applies special consideration to drug products intended for use in phase I clinical trials only. The FDA oversees drugs for use in phase I trials through its existing IND authority. Every IND must contain a section on chemistry, manufacture, and control of the investigational drug product (4). Even though the FDA is exempting some phase I drug products from compliance with some specific requirements of the cGMP regulations, the agency retains the ability to take appropriate actions to address manufacturing issues. Therefore, the pharmaceutical manufacturer must develop a strategy to ensure investigational drugs are produced under sufficient conditions to ensure the safety, identity, strength, purity, and quality of the drug are not adversely affected.

Following clinical trial manufacture, scale-up to commercial manufacture is a critical point in the development of a new product. Technology transfer is the term used for this final step in the transfer of the product from research and development to commercial production. Technology transfer (including scale-up) typically involves the manufacture of three consecutive, successful full-scale lots. Later in the chapter, it will be noted that cleaning validation data could be generated between these three lots manufactured in the commercial facility at full scale.

Each manufacturer must follow the regulations related to equipment cleaning and implement an equipment cleaning strategy that supports the new pharmaceutical product as it moves through the various stages of product development. The specific focus of this chapter will be applied to the requirements in late stage development (e.g., clinical trial material manufacturing) through commercial manufacturing (including the technology transfer between these two phases of pharmaceutical product development) as these stages are typically critically assessed during a PAI.

Equipment cleaning is an important component addressed during a PAI as is evidenced by the many companies that have received warning letters because of equipment cleaning violations referencing 21 CFR 211.67 (a–c) (5). Figure 2 provides a summary of all 483s issued from 2002 to June 2006 obtained via the website (http://www.fda.gov/) that reference violations to 21 CFR 211.67 (5). More specifically, 95% of the warning letters issued referencing 21 CFR 211.67 a–b relate to the firm's inability to produce written equipment cleaning procedures that are effective at eliminating the potential for product contamination.

To assist firms in complying with the Food Drug and Cosmetics Act, the FDA investigators comment on objectionable conditions and practices found after completion of an inspection in the form of the FDA 481. The FDA 483 is

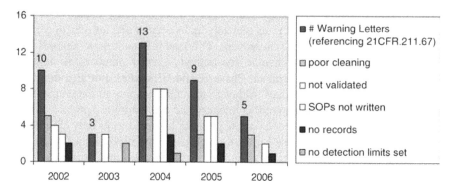

Figure 2 Summary of issued warning letters (2002–2006). *Abbreviation*: SOPs, standard operating procedures.

usually not intended to report favorable or acceptable condition that may have been observed during the inspection. A typical statement at the conclusion of an FDA 483 letter contains, but is not limited to, the following (5):

> The firm should notify the FDA office in writing, within 15 working days of the receipt of the 483 letter. This response should include the specific steps taken to correct the noted violations, including an explanation of each step being taken to prevent the recurrence of similar violations. If correction action cannot be completed within 15 working days, the firm must state the reason for delay and the time within which corrections will be completed. The 483 letter is not intended to be an all-inclusive list of deficiencies and it is clearly the responsibility of the pharmaceutical company to ensure that drug products are manufactured in compliance with appropriate regulations.

Some examples of the 483 observations referencing 21 CFR 211.67 include the following:

- Failure to clean and maintain equipment and utensils at appropriate intervals to prevent contamination that would alter the safety, identity, strength, quality, or purity of the drug product. Refer to 21 CFR 211.67 (a) Equipment and utensils shall be cleaned, maintained, and sanitized at appropriate intervals to prevent malfunctions or contamination that would alter the safety, identity, strength, quality, or purity of the drug product beyond the official or other established requirements.
- Written procedures are not adequately established for the cleaning and maintenance of equipment, including utensils, used in the manufacture, processing, packing, or holding of a drug product. Refer to 21 CFR 211.67 (b) Written procedures shall be established and followed for cleaning and

maintenance of equipment, including utensils, used in the manufacture, processing, packing, or holding of a drug product.

- Failure to keep records for the maintenance, cleaning and sanitizing, and inspection of equipment used in pharmaceutical production and failure to maintain a written record of equipment cleaning, maintenance, and use that includes the date and time of usage. Refer to 21 CFR 211.67 (c) Records shall be kept of maintenance, cleaning, sanitizing, and inspection as specified in 21 CFR 211.180 and 21 CFR 211.182.

A robust equipment-cleaning program will produce documented proof that can be presented during a PAI that a pharmaceutical manufacturer can consistently and effectively clean a system or a piece of equipment. It is important to note that development of a cleaning program can be rigorous and time consuming. Therefore, it is highly recommended that during development of the cleaning program, those closely involved should identify and correlate what *must be* achieved as required by regulations and what *can be* achieved given the time frame and resources available.

BACKGROUND

The objective of any equipment-cleaning program is to establish documented evidence that the cleaning process consistently provides a high degree of assurance that production equipment is free from materials that could contaminate subsequent products. A cleaning verification program or a cleaning validation program provides this documented evidence. Cleaning verification provides evidence that equipment and systems are free from materials that would contaminate or adulterate subsequent product lots. Cleaning verification is appropriate for the manufacture and packaging of materials used in clinical trials given the limited number of lots manufactured and packaged and the number of process changes that take place during the various phases of development.

An alternate approach to cleaning verification is cleaning validation. Cleaning validation confirms the effectiveness of a cleaning procedure and eliminates the need for routine testing. Cleaning validation is typically executed by demonstrating that the cleaning process utilized on multiple lots of the same material and process is adequate to reduce the level of active pharmaceutical ingredient below a certain safety limit. Because of the nature of pharmaceutical products in early stages of product development (e.g., clinical trial compounds), changes such as unpredictable lot sizes or modifications to the formulated drug product mean that three identical lots of a product/formulation manufactured by the same process may not be packaged until a later stage of development when the process has been locked and the size of the clinical trials have increased. It is not uncommon for a clinical trial packaging or manufacturing department to support 20 to 30 compounds at any given time. Therefore, performing cleaning validation on a single compound may not be feasible prior to a PAI, and cleaning verification must be followed in this case.

Many companies choose to write a document that outlines the facility strategy related to equipment cleaning (commonly termed CMP). This type of document addresses topics such as responsibilities, facilities, cleaning strategies, analytical strategies, and residue limit justifications. In addition to describing the basic cleaning strategy/approach followed in the facility, the basic elements of an equipment cleaning master plan include:

- Detailed cleaning procedure, including critical cleaning parameters, such as cleaning agents, used (including concentration and quantity); cleaning conditions for washing, rinsing, and drying (e.g., water type, time, temperature, volume); gowning requirements (when applicable); and disassembly and reassembly instructions required to perform a manual or an automated clean.
- Cleaning agents (other than water) for cleaning product contact surfaces that must be approved prior to use, and a toxicology opinion must be obtained.
- Instructions for conducting and documenting a visual inspection.
- Direct surface sampling (e.g., swabbing) instructions.
- Direct surface sampling location (e.g., swab locations) that are documented and justified, if applicable.
- Analytical methods for cleaning verification when performing direct surface sampling. The equipment surfaces contacted by the product (e.g., material of construction) must be considered for swab locations. If non-specific detection methods [e.g., total organic carbon (TOC), conductivity] for routine monitoring are used, the methods must be validated, and product surfaces should be considered.
- Equipment storage practices that allow for identification and storage of clean versus dirty equipment. Clean equipment must be stored in a manner to prevent contamination. A system must be in place to prevent dirty equipment, or equipment awaiting cleaning verification results from being used. Equipment cleaning and storage requirements must consider microbial and viral control strategies for the product as appropriate (e.g., biotech processes).
- Specific training required by personnel prior to performing equipment cleaning-related tasks.
- Equipment cleaning acceptance criteria that must be based upon knowledge of the materials under evaluation and be scientifically justifiable.

Creation of an equipment CMP that addresses these key topics will provide product development with the appropriate information needed to: develop manufacturing processes to transfer to commercial manufacture and generate information to submit for drug approval for market.

CLEANING STRATEGIES: CLEANING VERIFICATION VS. CLEANING VALIDATION

Cleaning verification is demonstrated on each manufactured lot (typically by submission of cleaning swabs to the analytical laboratories) by showing that the active pharmaceutical ingredient has been removed to a level below the pre-established acceptance limit. On the other hand, cleaning validation is an extensive, multi-functional program where the entire manufacturing process is evaluated, from the equipment that will be used to the analytical methods needed to evaluate cleaning effectiveness. Cleaning validation starts with the generation of a validation protocol. The validation protocol would include rationale with regard to cleaning agents and procedures, equipment, equipment-swabbing locations, safety-based acceptance limits, selection of products used to demonstrate validation, and validated analytical methodology. In addition, the cleaning process must be performed on three identical lots utilizing the same formulation and lot size. Because of these more stringent requirements, cleaning validation is usually reserved for commercial product manufacturing rather than product development/manufacturing activities.

When selecting representative products to validate a cleaning program, there are several approaches available. The most conservative approach would be to validate the cleaning procedure for all compounds manufactured. However, this approach is expensive and resource consuming and is typically not practical. Therefore, a manufacturing facility could select a representative range of similar products and processes to perform cleaning validation. This representative range of similar products and processes are commonly termed ''model compounds.'' Model compounds can be chosen on the basis of solubility, structure, potency, and/or compounds that otherwise pose a unique challenge to the cleaning process. Table 1 provides attributes associated with example model compounds. Following the table, the thought process to ensure that these compounds are representative is outlined.

The example model compounds shown in Table 1 have differences in solubility that represent a wide range of compounds. Compound A and Compound C are practically insoluble in water, HCl, and NaOH and slightly soluble in methanol. Compound B is sparingly soluble in water and HCl, practically insoluble in NaOH, and freely soluble in methanol. Compound D is practically insoluble in water, HCl, NaOH, and methanol. Compound E is soluble in water, practically insoluble in HCl, and freely soluble in NaOH and methanol. In addition to differences in solubility, the model compounds represent a wide range of manufactured strengths in either a tablet or capsule formulation (i.e., 0.5 to 50 mg for the smallest strength manufactured, and 0.75 to 200 mg for the highest strength manufactured) and a variety of chemical structures. The lowest and highest strengths manufactured are included in Table 1 to portray the wide range in manufactured strengths represented by the model compounds.

Table 1 Model Compounds

Compound name	Solubility	Lowest strength manufactured (mg)	Highest strength manufactured (mg)
Compound A	Practically insoluble[a] in water, HCl, and NaOH; slightly soluble[b] in methanol	25	200
Compound B	Sparingly soluble[c] in water and HCl; practically insoluble in NaOH; freely soluble[d] in methanol	2.5	100
Compound C	Practically insoluble in water, HCl, and NaOH; slightly soluble in methanol	2.5	20
Compound D	Practically insoluble in water, HCl, NaOH, and methanol	0.5	0.75
Compound E	Soluble[e] in water; practically insoluble in HCl; freely soluble in NaOH and methanol	50	200

[a]Practically insoluble = <0.1mg/mL
[b]Slightly soluble = 1 to 10 mg/mL
[c]Sparingly soluble = 10 to 33 mg/mL
[d]Freely soluble = 100 to 1000 mg/mL
[e]Soluble = 33 to 100 mg/mL
Solubility definitions consistent with USP (United States Pharmacopeia).

After selection of the model compounds, an experimental approach for validation may be performed by spiking equipment with a known amount of product, cleaning the equipment, and collecting data from direct surface sampling (i.e., swabbing). If enough data is gathered and deemed acceptable, a cleaning validation package may be assembled and routine swabbing is no longer required for products represented by the model compounds. The cleaning validation data are considered acceptable if the established cleaning acceptance limits are satisfied on each surface.

DETAILED CLEANING PROCEDURES

Equipment can be cleaned using various types of cleaning methods, including cleaning in place, cleaning out of place, and manual cleaning. Circulating cleaning solutions through the equipment performs cleaning in place. Cleaning out of place involves removing and cleaning equipment remotely (i.e., in a wash pit). Manual cleaning is often performed in pharmaceutical product development since equipment is typically small in size and portable. Some facilities are set up with sinks in each processing room, and equipment cleaning is performed in the process room to prevent potential cross-contamination by limiting hallways to only clean equipment.

A detailed cleaning procedure must be clearly defined, including critical cleaning parameters such as cleaning agents used (including concentration and quantity); cleaning conditions for washing, rinsing, and drying (e.g., water type, time, temperature, volume); gowning requirements (when applicable); and disassembly and reassembly instructions required to perform a manual or an automated clean.

APPROVED CLEANING AGENTS

When drug products are not soluble in water, a detergent (i.e., cleaning agent) is typically utilized in the cleaning process. Acidic or basic cleaning agents can also be utilized when aqueous solubility is not sufficient. For difficult to clean compounds, it is common practice to utilize a methanol wipe to provide added assurance that no residue is present. Cleaning agents for cleaning product contact surfaces must be approved by the quality unit prior to use and a toxicology opinion must be obtained to ensure use of product is acceptable for humans. To ensure that the approved cleaning agents do not themselves contribute a contamination risk, cleaning agent validation packages are required if routine monitoring for the cleaning agent is not performed. Cleaning validation performed for cleaning agents should reference the toxicology opinion to determine cleaning acceptance limits for the cleaning agent validation studies.

EQUIPMENT CLEANING ACCEPTANCE CRITERIA

How clean is it? The calculation for cleanliness for a piece of equipment (i.e., cleaning acceptance limit) is complicated and must meet the following criteria: practical, achievable, verifiable, and scientifically justifiable. More recently, equipment cleaning has become a topic in the forefront of discussion because of the development of highly active drug substances termed "potent compounds" (6). In the manufacture of carcinogenic, mutagenic, or other compounds deemed potent, containment is essential and validation of the cleaning procedure is a way of ensuring reduced levels (e.g., cleaning acceptance limit) of surface containment and improved containment. To understand the concept of containment, one should know the three different types of contamination: microbial, particulate, and cross-contamination. Contamination can be defined as the undesired introduction of impurities of a chemical or microbiological nature, or of foreign matter, into or onto a starting material or intermediate during production, sampling, packaging or repackaging, storage, or transport or the presence of any substance in a product that makes it impure, unclean, or unfit for use. The answer to "How clean is it?" must include clean enough ensuring reduced levels (e.g., cleaning acceptance limit) of surface containment and improved containment.

In calculating cleaning requirements during the early phases of development, companies in the industry may use formulas that are based on representative cleaning scenarios. A common practice observed in many companies today involves the use of two methods: the 1/1000th of the therapeutic level formula and the 10-ppm formula. The 1/1000th method assumes that pharmaceuticals are often considered to be nonactive at 0.1% of their normally prescribed dosages. The second method uses the 10-ppm limit historically used to calculate commercial manufacturing limits. This method allows the maximum carryover of product to be calculated using lot sizes and shared equipment surface area. Depending on data availability, cleaning criteria are frequently based on the formula that results in the smallest contamination level of products. The more active a drug substance (e.g., potent compounds), the lower the cleaning criteria, and the larger the effort and cost of maintaining product quality and personnel safety, making the cleaning and decontamination a serious concern and a major cost contributor in manufacturing. In this situation, many companies choose to have dedicated manufacturing equipment for these highly active drug substances.

VISUAL INSPECTION

Visual inspection provides a rapid assessment of equipment cleanliness and is used to verify that there are no areas within the processing equipment that contain residues that can be seen with the unaided eye. Visual inspections must be conducted on dry equipment and documented appropriately. An acceptable visual inspection is a pre-requisite to performing direct surface sampling (i.e., swabbing) and is performed prior to equipment use. In addition, an acceptable visual inspection can be used as the cleaning verification limit for surfaces with no reasonable possibility of contacting a drug product and for small equipment that can be fully disassembled and fully visually inspected. For all cleaning validation activities or when a visual clean alone is utilized to confirm equipment cleanliness, detailed visual inspection requirements (e.g., visual inspection locations, adequate lighting) and materials needed to perform the visual clean (e.g., flashlight, mirrors) will be documented in a local procedure.

DIRECT SURFACE SAMPLING

Given the nature of the equipment and the level to which it can be disassembled, direct surface swabbing can be an appropriate sampling technique for assessing equipment cleanliness. During development and verification of the swab sampling analytical method, selection of a suitable swabbing solvent occurs and will be specific for the active ingredient. Specifically, the product of interest should be soluble in the swabbing solvent.

The location and number of swabbing locations for both verification and validation activities can be dictated by one or more of the following: product contact surface area (e.g., larger components may require a larger number of swabs), energy dissipation (e.g., roller compaction or tablet pressing regions are more likely to accumulate product), composition of the product contact areas (i.e., polycarbonate vs. stainless steel), and cleaning difficulty (tight corners, bends, and hard to reach places result in hard to clean locations that should be swabbed).

Swabs are typically constructed of a polyester knit that will not leave behind fibers after swabbing and possesses minimal extractable materials. Typically, once equipment is completely dry and visually clean, the justified swab locations/surfaces are sampled. A general swabbing procedure may dictate 20 strokes executed in the horizontal direction; then the swab is flipped and 20 vertical strokes are performed. Swabs can be immersed in the appropriate sample solvent and stored for testing purposes.

ANALYTICAL STRATEGY/ANALYTICAL METHODOLOGIES RELATED TO EQUIPMENT CLEANING

The requirement to verify cleaning between manufacturing runs in a product development facility presents a unique challenge to the analytical chemist. Laboratory analysis of cleaning samples can be time and resource consuming in nature. Some labs use high-performance liquid chromatography (HPLC) for cleaning validation while others utilize TOC analysis. When selecting the appropriate analytical technique for cleaning sample analysis, the key parameters to consider are the acceptance limit required for the cleaning verification assay and the molar absorptivity of the molecule plus solubility, ionization, or oxidation potential.

The use of HPLC for cleaning sample analysis typically supports selective or specific methods designed for specific drug product ingredients. Derivative UV spectroscopy and enzymatic detection and titrations are also being utilized for selective cleaning methods. Another emerging alternative to HPLC analysis is ion mobility spectrometry, which has been successfully implemented for trace determination in security applications for many years (7). Selective techniques are typically used when the target analyte is known before method validation. HPLC is most commonly used in the pharmaceutical industry for selective assays to support cleaning swab analysis, potency of drug product, and/or detection of degradation products. The potency assay for the drug product is typically developed and validated prior to the requirement for cleaning swab analysis and separation of process impurities or potential degradation products is typically not required for the cleaning swab assay. Since the HPLC method requirements for swab analysis are less stringent, it is deemed appropriate to use the highly stringent potency HPLC method to support cleaning sample analysis.

TOC or conductivity can be utilized for cleaning sample analysis supporting nonselective methods. As opposed to selective or specific methods that detect a specific ingredient, nonselective methods detect presence of a blend of ingredients. For example, a nonselective assay can be used to determine if there is any deviation from the control (i.e., the blank in the same solution) and the main advantage of this approach is that a variety of residues may be detected, including cleaning agents, excipients, previously manufactured products, degradation products, and active pharmaceutical ingredients. Regulatory agencies may challenge nonselective methods because the exact, specific source of the contamination may remain unknown following this approach. There are large advantages to the TOC methodology as a nonselective technique since the technique provides good sensitivity, rapid analysis time, and the ability to detect all residual carbon—independent of the source.

EQUIPMENT STORAGE AREAS

Equipment should be cleaned and stored to prevent contamination or carryover of a material that would alter the quality of the final product. Storage practices must be in place that allow for identification and storage of clean versus dirty equipment. The specific requirement is that clean equipment must be stored in a manner to prevent contamination (2). A system must be in place to prevent dirty equipment or equipment awaiting cleaning verification results from being used. Equipment cleaning and storage requirements should also consider control for microbial and viral contamination strategies for the product, as appropriate (e.g., biotech processes).

PERSONNEL TRAINING

Critical activities that occur throughout the development of new pharmaceutical products should be described by procedures [e.g., standard operating procedures (SOPs)]. Documentation must exist to demonstrate that employees directly involved in manufacturing operations have been fully trained. A requirement specific to equipment cleaning is that individuals performing swab sampling should be appropriately trained to ensure that appropriate swab techniques are applied.

DISCOVERY TO EARLY FORMULATION DEVELOPMENT EQUIPMENT CLEANING STRATEGIES

Issues encountered during production of investigational drugs intended for use in small phase I clinical trials are different from issues presented by the production of drug products for use in the larger phase II and phase III clinical trials. Specifically, the requirement for fully validated manufacturing processes aren't always expected during phase I clinical trial material production. But, in a manufacturing facility where the number of new compounds entering the facility is limited, it may be possible to validate equipment cleaning for each compound as it is introduced into the facility. Or, a visual-residue limit might be successfully argued for an application with limited scope, such as introducing a new development compound into a development plan.

Lastly, there is always the approach to swab equipment, test the samples, and document the result. This option is resource intensive but very robust, which makes having the ability to look at a piece of equipment and sign a paper stating that it is ''visually clean'' an attractive option.

LATE DEVELOPMENT TO CLINICAL TRIAL MANUFACTURE EQUIPMENT CLEANING STRATEGIES

The number of new compounds entering a late phase development facility each year can be greater than the number of new compounds entering a production facility. As the compound goes through scale-up during development, the formulation may evolve, and the type of equipment used can also change. Therefore, it is not always practical to validate each compound, so following cleaning verification specific to each compound is an acceptable practice.

A robust process to evaluate equipment cleaning when manufacturing a new compound is needed until a full evaluation based on a completely developed formulation and cleaning procedure is possible. Since it is not typically possible to manufacture special batches for cleaning studies, cleaning studies can be built around the clinical compounds currently manufactured in the facility. Grouping and selecting one or more worst-case compounds can decrease the total work effort and also generate data to meet GMP regulations. Even when using a worst-case compound cleaning scenario, it is important to have identical cleaning procedures, cleaning agents, and equipment for use during the cleaning studies.

There are several accepted methods for selecting a worst-case product: most potent, least soluble, and/or hardest to clean. The general concept related to a worst-case scenario is that if one can clean the most potent and least soluble product adequately, then the same cleaning procedure should be adequate for lesser potent and more soluble products when cleaned by the same cleaning procedure. Equipment cleaning is of particular importance when facilities manufacture a large variety of products, such as a contract manufacturer.

The production of clinical trial material often falls within the domain of the research and development department and is often manufactured in a development facility. Therefore, the next product to be manufactured may not be known. One of the significant factors toward establishing cleaning acceptance limits is the batch size dosage of the following product. It is important to develop a strategy for development facilities to deal with this uncertainty, and this strategy may be to review the equipment history log and use the worst-case combination of parameters from previous manufacturing situations. Choose the smallest batch of product previously produced in order to use the largest daily dose of products made in the equipment for computing limits. Cleaning transfer could be conducted during batch scale-up to commercial manufacturing. This practice would require the resources to validate the cleaning of the new compound but would not affect the product timeline adversely. Traditionally, three successful runs of a standard process establish a validated process. Manufacture three separate batches followed by cleaning, swabbing, and testing for cleanliness.

COMMERCIAL MANUFACTURE EQUIPMENT CLEANING STRATEGIES

In some cases, companies do not perform formal cleaning validation until after the PAI because three full-scale lots have not yet been manufactured. In this situation, the cleaning process is usually validated concurrently with the validation of the manufacturing process. Up until this point, cleaning verification can be deployed by treating each manufacture and cleaning event as a unique entity.

SUMMARY

Proper documentation is the key to making a good first impression for any regulatory inspection to demonstrate a properly developed pharmaceutical product. Following are example documents to support a successful equipment-cleaning program in the event of a PAI:

- Approved CMP
- Approved SOPs for cleaning operations, including document training of personnel performing the work
- Approved analytical methods to support the equipment-cleaning program
- Justified sampling locations and cleaning acceptance limits
- Information supporting cleaning agents, including toxicology opinion and documented evidence of removal

Equipment cleaning is a critical element for a pharmaceutical manufactures, and is often a critical element that is evaluated by the FDA in a PAI. A robust equipment-cleaning program developed and executed concurrently with the development of the drug is essential, and will assist a facility in obtaining a successful outcome of the PAI. Throughout the chapter, equipment-cleaning requirements have been detailed and interpreted into equipment-cleaning strategies appropriate for pharmaceutical product development. As pharmaceutical product development continues to evolve, equipment-cleaning strategies will also evolve to meet the rigorous GMP requirements and to ensure patient safety.

ACKNOWLEDGMENTS

The author thanks Dr. Brian Pack for his contributions to the analytical methodology section of this chapter and Jeanette Buckwalter for her technical writing support.

REFERENCES

1. The Code of Federal Regulations. 21 CFR Part 211, Current Good Manufacturing Practice for Finished Pharmaceuticals.
2. Code of Federal Regulations. 21 CFR 211.67. 21 CFR Part 211, Current Good Manufacturing Practice for Finished Pharmaceuticals—Equipment Cleaning and Maintenance; Subpart D, Equipment.

3. ICH. Guidance to Industry: Q7A, Good Manufacturing Practice Guide for Active Pharmaceutical Ingredients.
4. Code of Federal Regulations. 21 CFR 312.23, 21 CFR Part 312, Investigational New Drug Application; Subpart B, IND content and format.
5. FDA. Center for Drug Evaluation and Research. Available at: http://www.fda.gov/.
6. Valvis II, Champion WL. Cleaning and decontamination of potent compounds in the pharmaceutical industry. Org Process Res Dev 1999; 3:44–52.
7. Alconox, Inc. Pharmaceutical Cleaning Validation References. Available at: http://www.alconox.com.

FURTHER READING

FDA. Guidance for Industry: Guideline on the Preparation of Investigational New Drug Products (Human and Animal). March 1991.

Forsyth R, Van Nostrand V. Using visible residue limits for introducing new compounds into a pharmaceutical research facility. Pharm Technol 2005; October 2:134–140.

Fourman GL, Mullen MV. Determining cleaning validation acceptance limits for pharmaceutical manufacturing operations. Pharm Technol 1993; 17:54–60.

Garvey C, Levy S, McLoughlin T. Cleaning validation: oral solid packaging equipment. Pharm Eng 1998; 18(4):20–24.

Gerber M, Perona D, Ray L. Equipment Cleaning in Clinical Trial Material Manufacturing and Packaging. Pharm Eng 2005; 25:46–54.

Hynes M. Preparing for FDA Pre-Approval Inspection, Drugs and Pharmaceutical Sciences. Vol 93. New York: Marcel Dekker, 1998.

Krull I, Swartz M. Cleaning validation. LC/GC Magazine 1999; 17(11):1016–1019.

Medina C. Compliance Handbook for Pharmaceutical, Medical Devices, and Biologics, Drugs and Pharmaceutical Sciences. Vol 136. New York: Marcel Dekker, 2003.

Meinertz Jensen H. Cleaning of a tablet packaging machine validated by a contractor: a case study. Pharm Eng 2004; 24(2):88–96.

Payne K, Fawber W, Faria J, et al. IMS for cleaning verification: the role of spectroscopy in process analytical technologies. Spectroscopy 2005; 20(1):24–27.

Ray L, Pack B, Wenzler L, et al. Equipment Cleaning Validation Modeling Approach for Clinical Trial Compounds. Pharm Eng 2006; 25:54–64.

Shea J, Shamrock WF, Abboud CA, et al. Validation of cleaning procedures for highly potent drugs. Pharm Dev Technol 1996; 1(1):69–75.

9

Conducting Stability Studies During Development to Ensure Successful Regulatory Approval

Julianne Eggert and Carol Fowler
Eli Lilly and Company, Indianapolis, Indiana, U.S.A.

BEGIN WITH THE END IN MIND

To quote Stephen Covey, author of the best-selling book *The Seven Habits of Highly Effective People* (1), when designing a product development stability program, you must "begin with the end in mind." The goal of a pharmaceutical product development program is to successfully register and launch a drug product in global markets. Passing a Food and Drug Administration (FDA) pre-approval inspection is a major milestone on the way to that goal. A robust and well-thought-out stability-testing program can help deliver that milestone.

The main purpose for running stability studies is to determine the expiration period and recommended storage conditions for the product. Stability data are also used to build a knowledge base describing the chemical and physical attributes of the product and what environmental factors (such as heat, water, oxygen, and light) can be harmful. Stability testing also helps guarantee that the patient receives the correct medicine. Designing the stability program to learn how those factors impact the product and how to best protect the product from those factors is the best way to drive to the goal.

REGULATORY GUIDANCE

To design a successful stability-testing program, there are many regulatory guidance documents that should be consulted. These guidance documents provide information on how to conduct a stability program to ensure that appropriate data are generated in support of a new drug substance or product.

Stability testing for product registration is one area covered by International Conference on Harmonisation (ICH) guidance documents. The ICH is a joint initiative involving both regulators and research-based industries from the European Union, United States, and Japan focusing on the technical requirements for medicinal products containing new drugs. This organization was initiated in the early 1990s and stability testing was one of the first topics to proceed through the stepwise process to recognition by the regulatory bodies from all three regions.

A series of guideline documents was developed to define the stability data required for registration of new drug substances and products within the ICH regions. The stability studies performed to support product registration should comply with these guidance documents. Currently, there are five guidance documents available. They begin with Q1A, Stability Testing of New Drug Substances and Products (2), which provides the basic protocol for stability studies for registration. For both new drug substances and new drug products, this document lists the number and types of batches, stability container closure systems, storage conditions, and time points that should be studied to support registration. It specifies that appropriate tests, analytical methods, and proposed acceptance criteria should be used, but references the ICH guidance documents on specifications and impurities for more detailed information. In addition to the general guidance on stability studies needed for registration, this document discusses stress testing of new drug substances and the required commitment to provide additional stability data on at least three production batches through the proposed retest period or shelf life, if not submitted in the original registration document.

The remaining four documents supplement this general protocol guidance. There is guidance Q1B, Photostability Testing of New Drug Substances and Products (3). This document provides instruction for carrying out photostability studies on new drug substances and drug products to demonstrate that light exposure does not negatively impact the materials. The testing outlined is performed on one batch from the registration stability study and is a stepwise approach with exposed drug substance, exposed drug product, drug product in its immediate package, and drug product in its marketed package, as necessary.

Guidance Q1C, Stability Testing of New Dosage Forms (4), was written to clarify the requirements for a new dosage form or line extension by the holder of the original submission. In this case, the requirements from Q1A are followed, but less data may be required at the time of submission.

In the parent guidance, Q1A, there is a mention of using bracketing or matrixing to reduce the amount of testing associated with the registration stability program. Guidance Q1D, Bracketing and Matrixing Designs for Stability Testing of Drug Substances and Drug Products (5), was written to provide more detailed guidance on the topic. This guideline discusses when each of these techniques may be considered and provides examples of them. It also discusses the potential risks with using these reduced testing designs.

The fifth guidance, Q1E, Evaluation of Stability Data (6), provides additional information on the evaluation and statistical analysis of the data generated following the Q1A guideline. This document provides a stepwise process for evaluating stability data and extrapolating that data to propose a retest period or shelf life. It discusses the application of linear regression, poolability tests, and statistical modeling to stability data for registration.

To supplement these guidance documents for the study of biotechnology products, an additional guidance was written. Q5C, Quality of Biotechnological Products: Stability Testing of Biotechnological/Biological Products (7), provides additional detail for the stability testing of biotechnological/biological products for product registration.

These guidance documents provide only the core requirements of the registration stability program. They do not provide all of the detail necessary to develop and manage the stability program in support of new product registration. Additionally, the abbreviated applications for registration of generic drug products is out of scope of the ICH documents but general principles may be taken from these guidance documents when studies to support registration are developed.

In the past, the FDA provided additional stability guidance in a document issued in 1987, Guideline for Submitting Documentation for the Stability of Human Drugs and Biologics (8). This document was followed by a draft FDA guidance issued in 1998, Guidance for Industry: Stability Testing of Drug Substances and Drug Products (9). In June 2006, the FDA withdrew both of these documents because of the agency's initiative, Pharmaceutical Current Good Manufacturing Practices (cGMPs) for the 21st Century (10,11). For stability guidance, the agency deferred to the ICH guidelines described previously.

The European Agency for the Evaluation of Medicinal Products Committee for Proprietary Medicinal Products (CPMP) has issued additional guidelines beyond their adoption of the ICH guidance. These guidance documents are as follows:

- Guidance CPMP/QWP/122/02, Stability Testing of Existing Active Substances and Related Finished Products (12), is an adaptation of the ICH guidance documents to be applied to those compounds for which a drug product has already been authorized within the European Community.

- Guidance CPMP/QWP/576/96, Stability Testing for Applications for Variations to a Marketing Authorisation (13), provides guidance on the stability data that should be generated in support of variations made to a marketing authorization. It provides some examples of variations and the types and amounts of stability data that would be expected to support them.
- Guidance CPMP/QWP/2934/99, In-Use Stability Testing of Human Medicinal Products (14), provides guidance on stability testing to establish the amount of time a multidose product can be used after it has first been opened.
- Guidance CPMP/QWP/159/96, Maximum Shelf-Life for Sterile Products for Human Use after First Opening or Following Reconstitution (15), states that studies should be conducted to support the practical use of sterile products. It also provides sample wording to include in user labeling specifying the appropriate hold times and storage conditions once the product is opened, diluted, or reconstituted.
- Guidance CPMP/QWP/609/96, Declaration of Storage Conditions: A: In the Product Information of Medicinal Products, B: For Active Substances (16), is an annex to the ICH stability guidance documents providing uniform storage condition statements for products. On the basis of the stability data generated following the ICH guidance, an appropriate storage condition labeling statement is suggested along with additional storage statements, where relevant.
- Guidance CPMP/QWP/072/96, Start of Shelf-Life of the Finished Dosage Form (17), outlines how to determine the expiration date of a drug product on the basis of the date of release or date of production.

The World Health Organization (WHO) has also issued guidance on the performance of stability studies. Guidelines for Stability Testing of Pharmaceutical Products Containing Well Established Drug Substances in Conventional Dosage Forms were issued as Annex 5 to the WHO Expert Committee on Specifications for Pharmaceutical Preparations Technical Report Series, No. 863, 1996 (18). This guidance was revised in 2003 and 2006 because of changes in the long-term storage condition to support climatic zone IV regions (19,20). A new stability guidance document to replace the 1996 guidance is in review.

As for other countries not mentioned specifically, many have adopted either the ICH or the WHO guidelines as the basis for their stability testing requirements.

PRODUCT DEVELOPMENT STUDIES TO SUPPORT REGISTRATION

Stress Testing and Screening Studies

Stability testing begins as part of the initial characterization of a potential new molecular entity. This early experimental work provides information on the chemical stability properties of the compound and is used to assess promising compounds for product development. It may also help in the selection of the best salt forms to advance to dosage form development.

Screening studies are performed to assess the stability of a compound in its solid state and in solution. Studies of the compound in its solid state are conducted by exposing it to increasing temperatures and moisture. Storage conditions of up to 70°C, with and without 75% relative humidity (RH), may be used for these tests. The compound may also be exposed to higher than normal levels of UV and visible light as part of these tests. Studies performed on the compound in solution include exposing it to various pH levels and to radical initiators or oxidizers. The testing at this stage is looking for major changes in chemical assay or properties. The data generated at this stage of development are considered preliminary and provide a basis for more comprehensive stress testing if the compound is chosen for clinical development.

As development of the new drug substance is continued further, more formal stress testing is carried out. Similar to the screening studies, this testing exposes the compound to various conditions in the solid state and in solution. These tests serve to determine degradation pathways and assure that the proposed analytical methods are suitable for measuring potential changes in the compound. These stress-testing studies need only be carried out on one lot of drug substance and should follow the guidance provided in the ICH Q1A guidance. This testing is conducted in the early phases of development. Results from these studies are included in the product registration document.

Clinical Trial Stability

During the course of product development, batches of material must be placed on stability to support the use of the material in clinical trials. These studies are conducted concurrently with the clinical trial to ensure the safety, identity, strength, purity, and quality of the material being administered to volunteers or patients. The data generated from these clinical trial stability studies may also be used to support the formulation and package development process. Stability data should be generated to support the long-term storage condition of the material, as well as temperature excursions that may be experienced during the trial. Clinical trial material is shipped to various locations around the world and may be stored under varying conditions. If the material is exposed to hotter or colder temperatures than what is recommended, data are needed to demonstrate under what conditions the material still meets acceptance criteria.

For a room temperature material, the testing protocol may be modeled on the ICH Q1A stability guidance. If the clinical trial is located in an ICH zone I or II region, long-term testing at 25°C/60% RH is likely adequate. If the clinical trial has investigator sites in zone III or IV regions, long-term testing at 30°C/65% RH may be necessary to support the higher temperatures and humidity in those regions. Accelerated testing at 40°C/75% RH for up to six months is routinely included in these protocols to support high temperature excursions during shipment and storage and to support extrapolation of the long-term data for the purposes of setting product shelf life. Low temperature studies may also be needed to support low temperature excursions that may be observed during shipping and distribution.

Other special studies to be described below may also be required for clinical trial materials. For example, if the clinical trial material will be diluted or reconstituted prior to use, stability data should be generated to support these activities and hold times.

Several other types of development studies are needed above and beyond the support of clinical trial materials to register a product. The number and types of studies needed depend on the dosage form and the shipping and distribution pathways for the product.

Bulk Drug Product Hold-Time Studies

Solid oral dosage forms manufactured and held in bulk for more than 30 days before packaging require stability data to support hold time. In cases where drug product manufacturing takes place at a different site than finishing, or a drug product is sold in a variety of packages, it is a good practice to have 12 months stability data to support the drug product storage in bulk. To conduct these hold-time studies, samples are stored in miniature simulators of the larger bulk drug product container. The simulators consist of a container/closure system composed of the same components as the actual container/closure system used to store the drug product, but are smaller in size and contain fewer dosage units. The simulator must provide equal or less protection from environmental conditions than the actual system.

These are one-time studies conducted during product development, most often during phase III, using the formulation planned for product registration.

Temperature Excursion/Cycling Studies

Temperature excursion/cycling studies are done to study the effects of short-term temperature excursions that a drug product may experience during shipping, distribution, and storage.

For early-phase clinical material, data from accelerated condition storage may be used to justify use of material that has experienced excursions that occur on the high-temperature side of the acceptable storage condition. If data are needed to justify temperature excursions below the low temperature of the storage conditions, an additional time point at a colder condition (refrigerated for a room temperature product and frozen for a refrigerated product) may also be included in the study. In the case of solid oral dosage forms, one would not expect colder temperatures to cause stability-related issues from a chemical standpoint, so at a minimum, an examination of the physical attributes of the

Table 1 Sample Short-Term Temperature Excursion Conditions

Product label storage condition	Controlled room temperature	Refrigerated	Frozen[a]
Stability storage conditions	−20°C for 2 days	−20°C for 2 days	25°C/60% RH for 2 days
	60°C/75% RH for 2 days[b]	40°C/75% RH for 2 days[b]	

[a]No short-term low temperature excursion study is required for frozen products.
[b]Pull samples after one day and hold them to test if acceptance criteria are not met after two days.

Table 2 Recommended Cycle Times

Product label storage condition	Controlled room temperature	Refrigerated	Frozen
Stability storage conditions	–20°C for 2 days followed by 40°C/75% RH for 2 days. Repeat for a total of 3 cycles	–20°C for 2 days followed by 25°C/60% RH for 2 days. Repeat for a total of 3 cycles	–20°C for 2 days followed by 5°C for 2 days. Repeat for a total of 3 cycles

Pull samples after one and two cycles and hold them to test if acceptance criteria are not met after three cycles.

drug product should be conducted. The impact of freezing on other dosage forms such as parenteral products, solutions, or bioproducts must also be considered. Some dosage forms no longer meet acceptance criteria upon freezing.

During later phases of development, more extensive studies are needed to support the product throughout the expected shipping/distribution chain of the product (21). More stressful conditions and temperature cycles are used to model potential excursions (Tables 1 and 2). Keep in mind that the goal is not to test the product until it fails acceptance criteria; rather it is to determine the extremes beyond which the product should not be used.

Verification of Stability During Shipping

Most companies are shipping drug products via a known route using qualified shipping methods or by using qualified packaging. A verification of stability during shipping can be done by either modeling the shipping routes or sending samples through the shipping process. Testing of critical chemical and physical parameters is conducted before and after shipping. Data from this study will demonstrate that drug product quality is maintained through changes of environmental conditions and physical handling during the shipping process.

Reconstitution Stability

Some drug products require mixing with a solution or reconstitution with a diluent prior to use. The stability of a drug product after reconstitution needs to be evaluated over the shelf life of the product. During product development, all desired diluents, dilution volumes, and storage conditions are to be examined. A reconstitution study is conducted similarly to the "dry state" (the product prior to mixing or reconstitution) study rather than carefully replicating patient use as is done in an in-use study, to be described below. The goal is to confirm the stability of the reconstituted drug product over its shelf life, using recommended diluents and diluent volumes, for the time period and storage conditions desired on the label. For immediate-use vials, the reconstituted shelf life could be as short as 8 to 24 hours. For multiuse cartridges and vials, the reconstituted shelf life could be much longer and be measured in days or weeks. The study on reconstituted drug product should include all stability-indicating chemical and physical tests, and be conducted at the beginning of the "dry state" study and annually over the course of the shelf life.

In-Use Studies

In-use stability studies are conducted on multidose products to determine the period of time the product can be used once the container is opened and the first dose removed. Studies are to be designed to simulate the actual planned use for the product. For example, products that are reconstituted before dilution into a bag for intravenous use must be studied both after reconstitution and after dilution to ensure product quality for expected usage periods. Is the vial reconstituted and held for hours or days before dilution? How long after dilution is it administered? Over what period of time is it administered? All of these parameters must be covered in the study design.

The product quality of injectables, suspensions, and ointments is to be evaluated to ensure that they can perform when exposed to parameters such as light and changes in temperature. Diluent compatibility studies are helpful in determining if any commonly used diluents should be exempted from use because of insolubility and excipient interaction issues. Diluents that are not compatible with the product and cannot be used should be listed on the label.

If multiple diluents are acceptable, in-use studies must be designed for each. Testing should be conducted at intervals comparable with expected practice (e.g., daily for a pen cartridge, at meal times for a diabetes care product). If the reconstituted product may be stored at multiple conditions (e.g., refrigerated or room temperature), an in-use study must be conducted for each condition.

Testing includes the chemical, physical, and microbial content of the product that have the potential to change over the prescribed usage period. Batches of drug products are to be tested for in-use stability over the desired product shelf life.

Container Orientation

Interactions between the container closure system and the product could impact quality and stability. Studies on liquid products are to be conducted with product stored in the upright and inverted/horizontal orientations to demonstrate whether product contact with the container closure system impacts physical, chemical, or microbial stability. Dry powders, lyophilized drugs, and other products requiring reconstitution call for an evaluation of both the upright and inverted orientations of the reconstituted product to ensure that product quality is not impacted by orientation. If no orientation-related differences are found during development or clinical studies, drug product primary stability batches may be placed on stability in one orientation.

PRIMARY STABILITY STUDIES FOR REGISTRATION

Primary stability studies for registration are the most important studies conducted during product development. Data from these studies are used to set specifications, and expiry or retest periods. ICH Q1A defines the stability data package sufficient for submitting to Europe, Japan, and the United States for a new drug substance and product for registration. The general principle states that the purpose of stability testing is to provide evidence for how the quality of a drug substance or drug product varies with time under the influence of a variety of environmental factors such as time, temperature, humidity, and light, and to establish a retest period for drug substance or a shelf life for a drug product and recommended storage conditions.

ICH Requirements

Basic ICH requirements are straightforward, such as

- batch selection,
- use of the proposed container/closure system for market,
- storage conditions, and
- testing frequency (time points).

The ICH introduces the concept of an intermediate condition, which for 25°C products, is set at 30°C/65% RH. This condition is designed to moderately increase the rate of chemical degradation or physical changes for drug substances or drug products intended to be stored long term at 25°C. Samples stored under these conditions are tested only if specification failures or significant changes are observed for samples stored under accelerated conditions. All tests scheduled at the accelerated conditions are conducted on the intermediate samples, and time points begin as soon as specification failures or significant changes are observed. Testing requirements for the intermediate condition for products to be stored in a refrigerator or freezer are also described.

Climatic Zones

The ICH is intended only for the registration of products in Europe, Japan, and the United States. Many other regions of the world have based their pharmaceutical regulatory requirements on the ICH. Because of differing climates, different parts of the world have been divided into four different climatic zones, with average temperatures and relative humidity values listed:

- Zone I: 20°C/42% RH (includes Russia, Canada, Northern/Eastern Europe)
- Zone II: 22°C/52% RH (includes United States, Southern Europe, Japan, Australia, China)
- Zone III: 28°C/35% RH (includes Chad, Jordan, much of Northern Africa)
- Zone IV: 27°C/76% RH (includes Brazil, India, Southeast Asia, much of Central America)

These average temperatures and relative humidity values translate into different long-term stability storage conditions for drug products. The ICH defined the long-term storage condition for zones I and II as 25°C/60% RH. Regulatory guidance from countries in zones III and IV require long-term stability data from samples stored under conditions more stressful than the 25°C/60% RH. Zone IV has split into two different camps. Some countries state that they are zone IV (a) and accept long-term stability data from the 30°C/65% RH condition. Zone IV (b), which includes the hotter and more humid areas such as Brazil and Southeast Asia are requesting long-term data from the 30°C/75% RH condition. Several country-specific and WHO guidance documents include two acceptable long-term conditions, 25°C/60% RH and 30°C/65% RH. The submitting firm is to select the condition appropriate for the climatic zone in which they wish to submit. Many of these guidance documents are adding footnotes to the long-term condition, stating that testing at higher humidity such as 30°C/75% RH is also acceptable.

PRE-APPROVAL INSPECTION READINESS

At inspection time, an investigator may ask for records and ask to speak with various employees as described below regarding the stability program. Inspections may be unannounced or scheduled in advance. Firms must always be inspection-ready. From a stability perspective, preparation in the areas below is necessary.

People

All employees working to support product registration are required to be regularly trained about cGMPs and have documented experience, education, and training to perform their specific job duties. Part of this training should include regulatory inspection readiness. From a stability management personnel stand-point, it is a good practice to specifically identify which people will meet with investigators to talk about stability management processes. If your department is large and the personnel managing the stability program are not the same as those conducting the testing or managing the facility and monitoring systems, identify the experts for each area to appropriately discuss their procedures and work. Do not try to answer questions outside your area of expertise. Make sure that the right people are prepared to respond to an investigator's questions and provide appropriate documentation.

Prepare specific employees to discuss and provide documentation regarding the following:

- The stability program, protocol, and study processes (Note that if the site of a pre-approval inspection differs from the site where the stability studies were conducted, have a process in place to physically or electronically transmit documents to the investigators. Consider whether experts will travel to the site of the inspection. Be prepared to host investigators at the site of the stability studies if they choose to delve more deeply into the stability section of the registration submission)
- The stability sample management and inventory control
- The stability data—laboratory processes, including
  - stability-indicating method development and validation,
  - testing procedures and laboratory sample management,
  - raw data documentation and handling, and
  - result reporting and verification.
- The stability facility, including
  - installation,
  - qualification, and
  - maintenance.
- Monitoring
  - Know how the chambers are monitored and alarmed and get the related documentation.
  - If the monitoring system is electronic, have a person available to talk about the computer system.
  - Describe the procedure for when a chamber's condition goes out of limits, including how you respond to excursions, and how are they documented and investigated.

Laboratory Records

Locate all laboratory records (notebooks, data binders, electronic record refer-ences, instrument files, test method documents) related to stability studies conducted on the product(s) subject to the inspection. Be prepared to produce copies of such records within one business day.

Facility Records

Gather or locate all records related to the stability facility (chamber, monitoring system, utilities, and backup systems). Include records documenting the installation, qualification, maintenance, and retirement (if applicable) of any equipment or instrumentation related to the facility. Be prepared to produce copies of such records within one business day.

Reports

Gather or locate any stability-related reports relevant to the inspection. Be prepared to produce copies of such records within one business day.

Investigations

Prior to the inspection, create a list of any stability-related deviation, suspect analytical result, and chamber condition excursion investigations impacting the product(s) subject to the inspection. Be prepared to discuss the details of the investigation and gather or locate the related documentation.

Procedures

Clear, concise, standard operation procedures (SOPs) define business practices and help insure a compliant, inspection-ready stability program. Below are the major good manufacturing practice (GMP)-required procedures, with a summary of what critical information should be described.

Stability Program

This is an overarching SOP that should tie together all of the other requirements and contains pointers to your other procedures, or describes how other procedures apply to the stability program. Topics included

- Scope of the stability program
  - Does the program cover only investigational products or marketed products?
  - Drug substance? Drug product? Intermediates? Devices?
  - How are batches placed on stability chosen?
  - What markets do you support?
- Management of deviations from the stability program
- Management of the changes to the stability program

- Protocol requirements and approval process
- Sample selection from production
- Stability study management and approval process
- Stability facility and equipment management
- Stability sample management and inventory control
- Laboratory testing of stability samples
- Monitoring of the stability program
- Record retention

Deviations

Firms must have a procedure describing the business process for discovering, investigating, documenting, reviewing, and approving deviations. Different levels of deviation may be described ranging from minor to major, with different levels of approval for different levels of deviations. Include requirements for corrective and preventative actions so that reoccurrences may be avoided.

Change Management

Changes to the stability program are inevitable, especially during product development. A process for documenting changes to stability protocols, studies, and the stability facility includes requirements for drafting the change proposal, conducting an impact analysis, review, approval, and closure.

Quality Management

Personnel are required to notify supervision of all issues related to the stability program that could adversely affect the product or patients taking the product. This SOP should describe the firm's processes for the appropriate notification to management and the quality unit of these potential issues. Management and quality unit are then responsible to ensure that issues are resolved or escalated appropriately.

Protocol and Study Management

This SOP should describe how protocols are written and studies are set up. Topics to include are:

- When protocols are required.
- Information to include in a protocol, such as
  - drug substance or drug product covered by the protocol,
  - storage conditions,
  - time points,
  - test methods,
  - specifications or acceptance criteria,

- expiry or retest period (if applicable), and
- rationale for protocol design.
- Detailed protocol approval process.
- Information to include in a stability study, such as
  - batch-specific information (batch ID, manufacturing date, packaging date, etc.),
  - reason for the study,
  - study start date,
  - protocol used,
  - laboratory responsible for testing,
  - number of samples required (include overage), and
  - orientation of samples.
- Detailed study approval process.
- Define study start date—Does your firm use the manufacturing date of the batch or the date the samples are placed in the stability chambers as the start date of the study? Choose one and document which one is used.
- Expiry time point—Depending on how you define study start date, ensure that a time point is scheduled within 30 days of the batch expiration or retest date.
- Changes to a study or protocol—Define what major changes are made to a study or protocol that require a formal change management process. At a minimum, the following changes are formally managed and approved by the quality unit:
  - Adding or deleting a time point or test
  - Making a change to a specification or acceptance criteria
  - Changing the laboratory responsible for running the test (to ensure they are qualified to run the test method)
  - Cancelling an ongoing study

Stability Facility Management

Many firms use either their analytical instrumentation or equipment SOPs to cover the stability chambers. Consider writing a procedure to give more details describing

- how chambers are identified and how access to the chambers is controlled,
- temperature and relative humidity control ranges,
- qualification requirements (IQ/OQ/PQ),
- preventive maintenance requirements,
- monitoring requirements and probe placement,
- requirements for changing a chamber storage condition,
- requirements for documenting a chamber out of limits and responding to alarms, and
- facility backup systems.

Stability Sample Management and Inventory Control

During the FDA pre-approval inspections, firms have been asked to reconcile physical inventory with inventory records. A comprehensive procedure from sample receipt in the stability area through sample disposition at the end of the study will help ensure a controlled stability sample inventory. Include requirements for the following:

- Documenting receipt of samples into the stability area
- Sample labels
- Storage of the samples in the chambers—Document the time, date, location, and orientation (upright, inverted, random, etc.) of the samples that are stored in the chamber
- Managing the sample pull—Define how you schedule pull dates and any acceptable ranges surrounding that date. Decide how you will manage pulls that are scheduled to occur on weekends and company holidays. Will you allow samples to be pulled early or late? How many days? How long is one month? Is it 30 days, or do you divide 365 by 12 and pull on that date?
- Chain of custody—Define how the samples will get to the testing laboratory. Do you deliver or do they pick up? If the samples are shipped off-site, describe the shipping process.
- Sample disposal—Define when samples can be discarded and whether you require reconciliation with your inventory system.

Laboratory Analysis of Stability Samples

Stability samples are often tested in the same laboratories that conduct batch release and development testing. Specific requirements for stability samples should be outlined in this SOP. These include the following:

- Storage of the sample in the laboratory prior to testing—How does the lab log in the sample? Where do they keep it? Are there any special handling requirements?
- Testing is conducted using stability-indicating methods
- Describe the required testing completion time for stability samples
- Define whether the lab may use batch release results for the initial (time zero) time point—Many firms allow this if the results are generated by the same laboratory using the stability test method within 30 days of sample storage in the chambers.
- Require laboratories to compare results with previous time point results to identify and investigate suspect analytical results—A separate procedure on conducting a suspect analytical result investigation is recommended.

Other

Many other required procedures impact the stability program, but are more general and not specific to stability. This is not an exhaustive list, but at a minimum, write SOPs describing

- use of LIMS, if applicable;
- use of stability chamber monitoring system, if applicable;
- good documentation practices;
- records retention;
- specifications;
- equipment;
- analytical test methods;
- analytical testing practices;
- sample management;
- investigation of suspect analytical results (out of trend/out of specification);
- reference standards; and
- management of an inspection from a regulatory authority.

STABILITY INSPECTION FINDINGS

In a review of the FDA inspection documents and warning letters, it is apparent that the stability testing of drug products is an area of great interest to the agency. Historically, the FDA has consistently ranked drug product stability in the top 10 reasons for drug recalls (22).

Inspection findings in the stability area cover the breadth of the function. They begin with the requirement for a *written* stability program that includes detailed stability protocols. These protocols must then be carried out as written. Several observations of samples not being tested at the established test intervals were made by the agency. The protocols must also include appropriate tests for the product, such as preservative effectiveness, throughout the shelf life for a preserved product or sterility at the end of the shelf life for a sterile product.

There are also the FDA Form 483 findings concerning the operation of the stability-testing chambers and the maintenance of stability samples. Documentation of the ability of the stability chambers to maintain the appropriate temperatures and humidity throughout the stability study has been requested during an inspection. A request for documentation of sample storage and removal from the chamber, as well as an inventory of stability chambers, has also been observed in the inspection reports.

Laboratory findings are also commonly reported during stability-related FDA inspections. A prevalent finding is around the handling of out-of-specification stability data. Observations include the lack of a procedure for handling suspect and out-of-specification results, lack of appropriate investigation of out-of-specification results, and failure to notify the regulatory agency of failing stability results. The ability of the laboratory to conduct the studies was also noted with comments around the adequacy of laboratory staffing and resources along with appropriate management and technical oversight of the laboratory.

CONCLUSION

The stability program is always an area of interest for a pre-approval inspection. Careful preparation, clear SOPs that are followed, thorough documentation, and training will put you on the right track for a successful inspection, which is necessary for approval to market and in turn can lead to a positive product decision.

REFERENCES

1. Covey SR. The Seven Habits of Highly Effective People, Free Press 1989.
2. ICH. Harmonised Tripartite Guideline. Stability Testing of New Drug Substances and Products Q1A (R2), February 2003.
3. ICH. Harmonised Guideline: Stability Testing: Photostability Testing of New Drug Substances and Products Q1B, November 1996.
4. ICH. Harmonised Guideline: Stability Testing: Requirements for New Dosage Forms Q1C, November 1996.
5. ICH. Harmonised Guideline: Bracketing and Matrixing Designs for Stability Testing of New Drug Substances and Products Q1D, February 2002.
6. ICH. Harmonised Guideline: Evaluation for Stability Data Q1E, February 2003.
7. ICH. Harmonised Guideline: Quality of Biotechnological Products: Stability Testing of Biotechnological/Biological Products Q5C, November 1995.
8. FDA. Guidance for Industry: Guideline for Submitting Documentation for the Stability of Human Drugs and Biologics, 1987.
9. FDA. Draft Guidance for Industry: Stability Testing of Drug Substances and Drug Products, 1998 [draft].
10. FDA. Pharmaceutical Current Good Manufacturing Practices (cGMP) for the 21st Century: A Risk-Based Approach. Available at: www.fda.gov/cder/gmp/.
11. FDA. Guidance for Industry: Quality Systems Approach to Pharmaceutical CGMP Regulations, September 2006.
12. EMEA Committee for Proprietary Medicinal Products (CPMP). Guideline on Stability Testing: Stability Testing of Existing Active Substances and Related Finished Products CPMP/QWP/122/02, March 2004.
13. EMEA Committee for Medicinal Products for Human Use (CHMP). Guideline on Stability Testing for Applications for Variations to a Marketing Authorisation CPMP/QWP/576/96, December 2005.
14. EMEA CPMP. Note for Guidance on In-Use Stability Testing of Human Medicinal Products CPMP/QWP/2934/99, September 2001.
15. EMEA CPMP. Note for Guidance on Maximum Shelf-Life for Sterile Products for Human Use after First Opening or Following Reconstitution CPMP/QWP/159/96, July 1998.

16. EMEA CPMP. Note for Guidance on Declaration of Storage Conditions: A: In the Product Information of Medicinal Products B: For Active Substances CPMP/QWP/609/96, October 2003.
17. EMEA CPMP. Note for Guidance on Start of Shelf-Life of the Finished Dosage Form CPMP/QWP/072/96, December 2001.
18. WHO. WHO Technical Report Series, No. 863, Annex 5 Guidelines for Stability Testing of Pharmaceutical Products Containing Well Established Drug Substances in Conventional Dosage Forms, 1996. Available at: www.who.int/medicines/areas/quality_safety/quality_assurance/regulatory_standards/en.
19. WHO. WHO Technical Report Series, No. 908, Item 11.1 WHO Guidelines for Stability Testing of Pharmaceutical Products Containing Well Established Drug Substances in Conventional Dosage Forms, 2003.
20. WHO. WHO Technical Report Series, No. 937, Item 10.1 Stability Testing Conditions. 2006.
21. Lucas, TI, Bishara, RH, Seevers, RH. A stability program for the distribution of drug products. Pharm Technol, 2004; 28(7):68–73.
22. FDA/CDER. CDER Report to the Nation: 2005. Improving Public Health through Human Drugs [and previous yearly reports].

10

Computer Systems Validation During the Drug Development Process in Anticipation of Pre-Approval Inspections

Ludwig Huber

Labcompliance, Oberkirch, Germany

INTRODUCTION

Computers are widely used when conducting studies and developing documentation for new drug applications. Proper functioning and performance of software and computer systems play a major role in obtaining consistency, reliability, and accuracy of data. Therefore, computer system validation (CSV) should be part of any good development practice. It is also requested by the Food and Drug Administration (FDA) regulations and guidelines through the overall requirement that the "equipment must be suitable for its intended use."

There are no specific regulations or guidance documents from the FDA or international agencies related to using computers in pharmaceutical drug development. The FDA's Compliance Program Manual on Pre-approval Inspections/Investigations (1) refers to the Inspection Guide of Pharmaceutical Quality Control Laboratories (2) for inspection criteria when covering laboratory operations, including the use of computer systems.

The FDA has developed several specific guidance documents on using computers for other FDA-regulated areas. Most detailed is the Industry Guide:

General Principal of Software Validation (3). It deals with the development and validation of software used in medical devices. More recently, the FDA has released a guidance about using computers in clinical studies (4). This guidance states the FDA's expectations related to computer systems, and to electronic records generated during clinical studies.

Specific requirements for computers and electronic records and signatures are defined in the FDA's regulations 21 CFR Part II on Electronic Records and Signatures (5). This regulation applies to all FDA-regulated areas and has specific requirements to ensure trustworthy, integrity, and reliability of records generated, evaluated, transmitted, and archived by computer systems. Section 11.10(a) requires validation of computer systems used to create, modify, maintain, or transmit electronic records to ensure accuracy, reliability, and consistent intended performance.

By far the most detailed and most specific document that has ever been developed on using computers in regulated areas is the "Good Practices Guide on Using Computers in GxP Environments" (6). It has been developed by inspectors for inspectors of the Pharmaceutical Inspection Convention Scheme but is also quite useful for the industry. It has more than 50 pages and includes a six-page checklist recommended for use by inspectors.

Because of their importance, several industry organizations and private authors have already addressed computer validation issues such as the following:

- The Good Automated Manufacturing Practices Forum has developed guidelines for computer validation (7).
- Huber has published a validation reference book for the validation of computerized analytical and networked systems (8).
- The Parenteral Drug Association has developed a technical paper on the validation of laboratory data acquisition system (9).

All these guidelines and publications follow a couple of principles.

- Validation of computer systems is not a one-time event. It starts with the definition of the product, or project, and setting user-requirement specifications, and covers the vendor selection process, installation, initial operation, going use, and changes control and system retirement.
- All publications refer to some kind of life cycle model with a formal change-control procedure being an important part of the whole process.
- There are no detailed instructions on what should be tested. All guidelines refer to the intended use and to risk assessment for the extent of validation.

While in the past computer validation was focused on functions of computers used by single users, recently the focus is on network infrastructure, networked systems, and on security, authenticity, and integrity of data acquired and evaluated by computer systems.

SCOPE OF THE CHAPTER

This chapter will guide lab managers, IT personnel, QA personnel, and users of computer hardware and software through the entire validation process from writing specifications and vendor qualification to installation and the initial and ongoing operation.

It covers the following:

- Validation of computer systems typically used in pharmaceutical development. This includes, computerized analytical systems, laboratory data systems, laboratory information management systems, and document management systems
- Qualification of network infrastructure and validation of networked systems
- Validation of spreadsheet applications such as Excel™ worksheets
- Validation of functions that are required to meet electronic records and signature compliance, such as 21 CFR Part 11
- Documentation as required by regulations

Risk-based validation is a modern concept to optimize validation efforts toward computer systems that have a high impact on product quality. This optimization is especially important since the FDA has been using and supporting the risk-based approaches for compliance as part of the 21st Century Drug cGMP initiative (10). It requires two steps: (1) Define the risk of the computer system, e.g., high, medium, or low. (2) Define validation steps or tasks for each category. Huber (11) has described the principles and applications of risk-based computer validation. In this chapter, risk assessment and risk-based validation will be discussed when appropriate, together with the validation phases.

A lot of up-to-date information can also be found on and downloaded from the Internet, for example, http://www.labcompliance.com/tutorial/csv, a Web site dedicated to computer validation in laboratories with frequent updates on new regulations and guidelines.

VALIDATION OVERVIEW

Computer validation covers the entire life of a product. It starts when somebody has an idea about a product and ends when the equipment is retired. CSV ends when all records generated on the computer system have been migrated on a new one and validated for accuracy and completeness. Because of the length of time and complexity, the process has been broken down into shorter phases: design qualification (DQ), installation qualification (IQ), operational qualification (OQ), and performance qualification (PQ) (8). The process is illustrated in Figure 1.

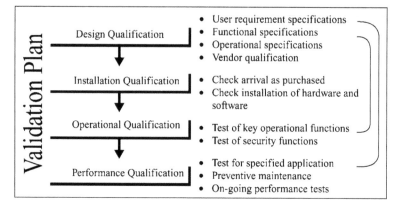

Figure 1 Computer validation phases. *Source*: From Ref. 8.

The approach a company takes should be described in a validation master plan that provides a framework for thorough and consistent validation. Regulatory agencies typically do not specifically demand a validation master plan. However, inspectors want to know what the company's approach toward validation is. The validation master plan is an ideal tool to communicate this approach both internally and to inspectors. It also ensures consistent implementation of validation practices and makes validation activities much more efficient. In case there are any questions as to why things have been done or not done, the validation master plan should give the answer. It should also include templates and examples that help validation professionals to validate individual systems.

Within an organization, a validation master plan can be developed for

- multiple sites,
- single sites,
- single locations,
- single system categories, and
- department categories, e.g., for development departments.

Computer validation master plans should include the following:

1. Introduction with a scope of the plan, e.g., sites, systems, and processes
2. Responsibilities
3. Related documents, e.g., validation policy
4. Products/processes to be validated and/or qualified
5. Validation approach
6. Risk assessment with examples of risk categories and recommended validation tasks for different categories
7. Steps for CSV with examples on type and extent of testing

8. Handling of existing computer systems
9. Validation of macros and spreadsheet calculations
10. Change of control procedures and templates
11. Contingency planning and disaster recovery
12. System obsolescence and removal
13. Training plans (e.g., system operation, compliance)
14. Templates and references to standard operating procedures (SOPs)
15. Glossary

For larger projects, a detailed individual validation project plan should be developed. An example would be implementing a laboratory information management system or networked chromatographic data system. This plan is derived from the validation master plan. It formalizes qualification and validation and outlines what is to be done in order to get a specific system into compliance. For inspectors it is a first indication on which control a department has over a specific computer system, and it also gives a first impression of the validation thoroughness.

A validation project plan should include sections on the following:

- Scope of the system—what it includes, and what it doesn't include
- System description
- Responsibilities
- Risk assessment
- Risk based test strategy and approach for DQ, IQ, OQ, and PQ
- Ongoing validation
- System retirement
- Time line and deliverables for each phase

DQ

The first step within a computer validation lifecycle is DQ. It "defines the functional and operational specifications of the instrument and details the conscious decisions in the selection of the supplier" (8). DQ should ensure that computer systems have all the necessary functions and performance criteria that will enable them to be successfully implemented for the intended application and to meet business requirements. Errors in DQ can have a tremendous technical and business impact, and, therefore, sufficient amount of time and resources should be invested in the DQ phase. For example, setting wrong functional specifications (FS) can substantially increase the workload for OQ testing, and selecting a vendor with insufficient support capability can decrease instrument uptime with a negative business impact.

Table 1 lists the recommended steps that should be considered for inclusion in a design qualification.

Table 1 Steps in Design Qualification

Description of the task the computer system is expected to perform
Description of the intended use of the system
Description of the intended environment (includes network environment)
Preliminary selection of the SRS and vendor
Vendor assessment
Final selection of the SRS
Final selection and supplier
Development and documentation of final system specifications

Abbreviation: SRS, system requirement specifications.

To set the system requirement specifications (SRS), the vendor's specification sheets can be used as guidelines. However, it is not recommended to simply write up the vendor's specifications because compliance to the specifications must be verified later on in the process during OQ and PQ. Specifying too many functions will significantly increase the workload for OQ.

Vendor assessment should answer the question: "How do you know that the system has been developed and validated in a quality assurance environment?" Depending on the risk and impact on (drug) product quality, answers can be derived from the following:

1. Documentation of experience with the vendor and product
2. Internal and external references
3. Assessment checklists (mail audits)
4. Third party audits
5. Direct vendor audits.

Assessment costs increase from one to five, and the final procedure should be based on justified and documented risk assessment. Two factors should be considered: (1) the risk the computer system has on (drug) product quality, and (2) the vendor risk. Criteria for the vendor risk include size of the company, company history, and representation in the (bio)pharmaceutical industry, and the experience with vendor. A vendor audit is recommended only if the product risk and the vendor risk are high, otherwise documenting experience and well-answered assessment checklists are sufficient.

IQ

IQ establishes that the computer system is received as designed and specified, that it is properly installed in the selected environment, and that this environment is suitable for the operation and use of the instrument. Table 2 lists steps recommended before and during installation.

Table 2 Steps Before and During Installation

Before installation

Obtain manufacturer's recommendations for installation site requirements.

Check the site for the fulfillment of the manufacturer's recommendations (space, utilities such as electricity, and environmental conditions such as humidity, temperature, vibration level, and dust).

During installation

Compare computer hardware and software, as received, with purchase order (including software, accessories, spare parts).

Check documentation for completeness (operating manuals, maintenance instructions, SOPs for testing, safety, and validation certificates).

Check computer hardware and peripherals for any damage.

Install hardware (computer, peripherals, network devices, cables) Install software on computer following the manufacturer's recommendation.

Verify correct software installation, e.g., are all files accurately copied on the computer hard disk. Utilities to do this should be included in the software itself.

Make backup copy of software.

Configure peripherals, e.g., printers and equipment modules.

Identify and make a list with a description of all hardware, include drawings where appropriate, e.g., for networked data systems.

Make a list with a description of all software installed on the computer.

Store configuration settings either electronically or on paper.

List equipment manuals and SOPs.

Prepare an installation report.

Abbreviation: SOPs, standard operating procedures.

OQ

"Operational qualification (OQ) is the process of demonstrating that a computer system will function according to its functional specifications in the selected environment" (8).

Before OQ testing is done, one should always consider what the computer system will be used for. There must a clear link between testing, as part of OQ, and requirement specifications, as developed in DQ phase. Testing may be quite extensive if the computer system is complex and if there is little or no information from the supplier on what tests have been performed at the supplier's site. Most extensive tests are necessary if the system has been developed for a specific user. In this case the user should test all functions. For commercial off-the-shelf systems that come with a validation certificate, tests should be done only of functions that are highly critical for the operation, or of those that can be influenced by the environment. Examples are data acquisition over a relatively long distance from analytical instruments at high acquisition rate.

Test number:
Specification:
Purpose of Test:
Test Environment (PC hardware, peripherals, interfaces, operating system, Excel version, service pack):
Test execution: Step 1: Step 2: Step 3:
Expected result: Acceptance criterion: Actual result: Comment:
Criticality of test: Low 0   Medium 0   High 0
Test Person: Printed name:_____ Signature:_____ Date:_____

Figure 2   An example of a test script. *Source*: From Ref. 12.

Proper functioning of back-up and recovery and security functions like access control to the computer system and to data should also be tested. Full OQ test should be performed before the system is used initially and at regular intervals, e.g., for chromatographic data systems about once a year and after major system updates. Partial OQ tests should be performed after minor system updates.

Tests should be quantitative. This quantification means inspectors would not only expect a test protocol with test items and pass/fail information but also expected results, acceptance criteria, and actual results. An example for a test protocol template is shown in Figure 2.

The extent of testing depends on the risk the system has on (drug) product quality and consumer safety. While a word processing system used to generate an SOP does not require any functional testing, a system used to test stability, or a drug needs extensive testing of critical functions.

PQ

"Performance Qualification (PQ) is the process of demonstrating that a system consistently performs according to a specification appropriate for its routine use'' (8). Important here is the word 'consistently.' Important for consistent computer system performance is regular preventive maintenance, e.g., removal of temporary files, regular virus checks for online systems, making changes to a system in a controlled manner, and regular testing.

In practice, PQ can mean testing the system with the entire application. For a computerized analytical system this can mean, for example, running system suitability testing, where critical key system performance characteristics are measured and compared with documented, preset limits. For example, a well-characterized chemical standard can be processed through the entire system five or six times, and the standard deviation of amounts is then compared with a predefined value. The use of analytical quality control sample with a comparison of the actual result and the known amounts is also a useful system check. If the computer system generates the results as expected, this also means that the software works 'as intended.' Such tests should be built into regular routine work. If such complete system tests are not possible or not practical, alternatively well-characterized data sets with expected outcome can be processed through the system, and the output should be compared with the output of the initial tests. Test cases and test conditions should be selected such that all key functions are tested. Such testing is called regression testing, and is also used to test overall functioning of the system after system updates.

VALIDATION OF MACROS AND SPREADSHEET APPLICATIONS

Macros programs are popular in pharmaceutical development for evaluating analytical data and deriving characteristics of the analyzed products. Spreadsheets are widely used in laboratories, e.g., data manipulation and report generation. Databases are used to correlate data from a single sample analyzed on different instruments and to obtain long-term statistical information for a single sample type. The processes may be automated using VBA scripts; for example, enabling the data to be transferred, evaluated, and reported automatically. In all of these programs, analytical data are converted using mathematical formulae.

Today's common understanding is that the programs themselves don't have to be validated by the user, e.g., MS Excel™. What should be validated are the calculations and program steps a user has defined and written. There should be good documentation on what the application program written by the user as an add-on to the core software is supposed to do, who defined and entered the formulae, and what the formulae are.

Validation activities should follow a life cycle model starting with defining user requirements, followed by design specification, code development, structural, and functional testing. For small projects, some of the typical phases like functional testing and testing in a user's environment can be combined (Fig. 3).

Most important is that changes should follow standard procedures for initiation, authorization, implementation, testing, and documentation. All activities

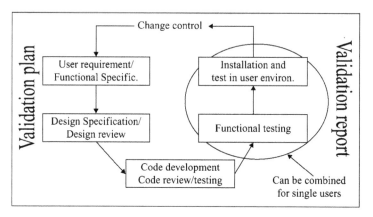

Figure 3 Life cycle of macros and spreadsheet programs. *Source*: From Ref. 12.

should be planned in the validation project plan and documented in the validation report.

The programmer and, later on, anticipated users should test and verify the functioning of the program. A frequently asked question is how much testing should be conducted. Testing should demonstrate that the system is providing accurate, precise, and reliable results. A spreadsheet, for example, may be tested using typical examples throughout the typical operating range. Results obtained by the program should be compared with results obtained by a calculator. If the spreadsheet is to be used to further evaluate small numbers, the test program should also include small numbers, or if an accurate calculation requires three digits after the decimal, the test should also use numbers with at least three digits after the decimal. The test program should also include stress testing with values outside the normal operation range.

Test cases should be linked and traceable to user requirements and FS in a traceability matrix.

Test protocols should be used with information on

- functions that should be tested,
- steps to be performed,
- equipment used,
- data inputs and expected output,
- acceptance limit,
- name of test person, and
- date of testing.

The tests should be repeated thereafter at regular or non-regular intervals. Therefore, any test case should be designed for reuse with using sets of data inputs and known outputs. If the program is used infrequently, it is a good practice to run the tests each time the program is used.

More details on design, validation, and use of macros and spreadsheet applications in GxP and Part 11 environments can be found in Huber (12).

CONFIGURATION MANAGEMENT AND CHANGE CONTROL

Any changes to specifications, programming codes, or computer hardware should follow written procedures and be documented. Changes may be initiated because errors have been found in the program, or because additional or different software functions or hardware may be desirable. Requests for changes should be submitted by users, and authorized by the user's supervisor or department manager. For initiation, authorization, and documentation of changes forms should be used. An example is shown in Figure 4.

After any change, the program should be tested. Full testing should be done for the part of the program that has been changed, and regression testing should be done for the entire program.

Form ID:			
System ID:			
System Location:			
Change Initiator:			
Description of change (should include reason for change and business benefit):			
Expected impact on validation:			
Authorization to change:	Name:	Signature:	Date:
Change implemented on:	Date:		
Comments (implementation testing):	e.g., document any observation and new version or revision number, and types of tests that have been performed.		
Completed by:	Name:	Signature:	Date:
Approved by:	Name:	Signature:	Date:

Figure 4 Template for change control.

QUALIFICATION OF NETWORK INFRASTRUCTURE AND VALIDATION OF NETWORKED SYSTEMS

Networked systems are used increasingly in pharmaceutical industry. These are computerized systems and, as such, must be qualified and validated to demonstrate suitability of their intended use. While validation of stand-alone computer systems is well described (8) and understood, there is still uncertainty on how to qualify network infrastructure and validate networked systems. On the other hand, inspectors are looking more and more into such systems, and validation of such systems is important for business reasons. For example, missing data from a bioequivalence project can be a disaster for the company and for individuals. The same holds for FDA submission delays caused by network failures.

Crosson et al. (13) and Huber (14) have addressed the quality assurance of networks. They recommended that a network be qualified as if it were a piece of equipment and then managed in a documented state of control.

Figure 5 shows a diagram of a client/server networked system connecting client computers in a laboratory and office computers to a server located in a computer room. The computer room also hosts mail servers. The laboratory computers with data system applications software control equipment with built in local area network cards and acquire data using TCP/IP protocols. Application software on these client computers is also used for data evaluation. Computers are connected to server computer through a hub. The server has relational database from Oracle with customized applications for data management, control

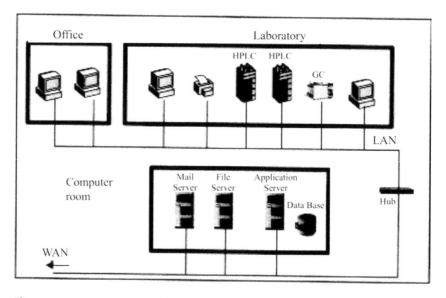

Figure 5 Example for a client/server networked system. *Source*: From Ref. 14.

charting, and other statistical evaluation, for generating electronic signatures and for review, backup, archiving, and retrieval of data.

Validating networked systems requires qualifying individual network components, for example, network devices. It also means qualifying authorized access to the networked system and qualifying data transfer between two computers, which qualify the interfaces of components at both sites. Validating the whole system requires running typical applications under normal and high load conditions and verifying correct functions and performance with previously specified criteria. For both qualification of the components and validation of complete systems, it is important to define a validation box. This description should well define which parts of the complete network need to be qualified and which are not affected. In the example as shown in Figure 5, the validation box for the laboratory data system includes lab computers, the file server, the applications server, and the database. Limiting the network qualification tasks to what components are used by network applications can save time in testing.

In the following paragraphs, the two steps for complete network qualification and system validation are briefly described. They have been extracted from Huber. (14)

Steps to Build a Qualified Network Infrastructure

- Specify network requirements. Specifications should include: network devices, software, computer hardware, and computer peripherals cables. Specifications are based on anticipated current and future use of the network.
- Develop a network infrastructure plan.
- Design network infrastructure and drawings.
- Select equipment and vendors for computers, network operating systems, network devices, etc.
- Order equipment: computer hardware, software (OS, NOS), network devices, and peripherals.
- Install all hardware devices according to design drawings and vendor documentation.
- Perform self-diagnostics, document hardware installation and settings (this completes the IQ part).
- Document this as network baseline.
- Make a backup of installed software and network configurations. Whatever happens, it should be possible to return to this point.
- Test communication between networked computers and peripherals, and access control, including remote access control.
- Develop and implement rigorous configuration management and change control procedure for all your network hardware and software. This task should also include updates of system drawings if there are any changes.

- Apply any system changes to a production environment, after verifying them in a test environment to insure that one does not impact the intended functionality of the system.
- Monitor ongoing network traffic using a network health monitoring software.

Steps to Define, Implement, and Validate Applications Running on a Qualified Network

- Develop a validation plan and schedule using your validation master plan as guideline.
- Specify application software that runs on the qualified network, e.g., networked data system.
- Select and qualify the vendor.
- Install application software and perform IQ (set up, define, and document configuration settings, verify 'proper' software installation through installation verification routines).
- Verify correct functioning of this application software. Apply common computer validation practices for this.
- Include network transactions under normal and high load in the testing. For this test you can decide to refer to verifications done as built in TCP/IP transfer protocol. The advantage is that this is built into the system and done on an ongoing basis. However, this is not 100% accurate, as there are rare situations where the test does not work. To be on the safe side, you should use something like MD5 hash calculation routines based on 128-bit strings. Ask the vendor of your application software; sometimes such tests are built in.
- Monitor on-going performance of your application. Type of performance test depends on the application.
- Monitor network connections and traffic using a network health monitoring software for networks supporting high-risk applications.

As part of the installation system drawings and diagrams should be generated. They are an absolute must not only for setting up a network and networked systems but even more important for maintaining them. They should be part of the IQ documents, and should include

- physical diagrams such as component locations and cabling, and
- logical diagrams like TCP/IP schemes, and how components interrelate with each other. If a dynamic IP addressing used in the network scheme, a procedure should also be in place to indicate how this dynamic IP addressing is being utilized (including the procedure for sub-masking of the IP addresses). This procedure will enable appropriate tracing of data or traffic flow, in case such a tracing is needed to prove the data integrity and security.

Networks change frequently, so maintaining these diagrams with documented version control is important. A good recommendation is to have procedures in place for regular review of these diagrams, for example, quarterly.

VALIDATION COMPLIANCE WITH ELECTRONIC RECORDS AND SIGNATURES

In 1997 the FDA issued a regulation that provides criteria for acceptance of electronic records, electronic signatures, and handwritten signatures (5). This provision was made in response to requests from the industry. With this regulation, entitled Rule 21 CFR Part 11, electronic records can be equivalent to paper records and handwritten signatures. The rule applies to all industry segments regulated by the FDA that includes good laboratory practice, good clinical practice, and current good manufacturing practice.

The use of electronic records is expected to be more cost effective for the industry and the FDA. The approval process is expected to be shorter and access to documentation will be faster and more productive.

The primary requirements of the new regulation for analytical laboratories are

- use of validated existing and new equipment and computer systems,
- secure retention of electronic records to instantly reconstruct the analysis,
- user independent computer generated time-stamped audit trails,
- system and data security, data integrity, and confidentiality through limited authorized system access,
- use of secure electronic signatures for closed and open systems, and
- use of digital signatures for open systems.

It is out of the scope of this chapter to discuss and explain how the requirements can be implemented in analytical laboratories. It has been described in an article series published in Journal of GxP Compliance (15). This chapter will elaborate only the validation aspect of the rule.

Part 11 requires that computer systems used to acquire, evaluate, and transmit and store electronic records should be validated. This is nothing new, and processes and steps to validate such systems should follow steps described earlier on in this chapter. Specific functions to be tested as required by Part 11 are

- limited authorized access to data and systems,
- computer-generated audit trail,
- binding signatures to records,
- accurate copies of electronic records, and
- ready retrieval for processing of data.

Required steps to achieve validation compliance are not different from other validation steps:

- Specify requirements, and include them in your user requirement specifications (URS) document.
- Develop test procedures to very that the functions meet the requirements.

Recommended test procedures include the following:

- Limited and authorized system access. This can be achieved by entering correct and incorrect password combination and verify if the system behaves as intended.
- Limited access to selected tasks and permissions. This can be achieved by trying to get access to tasks as permitted by the administrator and verify the system behaves as specified.
- Computer-generated audit trail. Perform actions that should go into the e-audit trail according to specifications. Record the actions manually compare and compare the recordings with computer-generated audit trail.
- Accurate and complete copies. Calculate results from raw data using a defined set of evaluation parameters (e.g., chromatographic integrator events, calibration tables, etc.). Save raw data, final results, and evaluation parameters on a storage device. Switch off the computer. Switch it on again and perform the same tasks as before using data stored on the storage device. Results should be the same as for the original evaluation.
- Binding signatures with records. Sign a data file electronically. Check the system design and verify that there is a clear link between the electronic signature and the data file. For example, the link should include the printed name or a clear reference to the person who signed, the date and time, and the meaning of the signature.

DOCUMENTATION

On completion of CSV, documentation should be available that consists of the following:

- Validation master plan
- Validation project plan
- Design qualification document consisting of
  - URS
  - FS (URS and FS can be combined into one SRS document)
- Vendor assessment
- Installation qualification document (for computer systems this document should include an installation verification test report)

- Network diagrams (for networked systems)
- Entries of instrument ill in the laboratory's instrument data base
- Test plan with test traceability matrix
- Procedures for testing with expected test results and acceptance criteria
- Qualification test reports with signatures and dates
- List with authorized users (the list should be signed and regularly reviewed)
- Procedures for preventive maintenance
- Change control procedures and change logs
- PQ test procedures and representative results
- Validation summary report

CONCLUSION

Most important for the entire process of CSV in pharmaceutical development is proper planning, execution of validation steps according to the plan, and documentation of the results. The process should start with the definition of the use of the computer system, and the development of user requirement and functional specifications. For computer systems a formal vendor assessment should be made. This assessment can be made through checklists, vendor documentation combined with internal and/or external references, or for very complex systems through a vendor audit.

The installation should be documented. The accuracy of software installation should be verified and for networked systems drawings with diagrams should be generated. The instrument should be tested for compliance to user requirement and functional specifications, as defined during the design qualification. Critical parameters should be tested before and during routine analysis. System suitability testing and the analysis of quality control samples, together with additional tests of critical parameters, are suitable tools for this on-going performance qualification.

All changes to the system should follow a documented change control procedure. They should be authorized before implementation, and a formal assessment should be made for the need and extent of revalidation.

REFERENCES

1. FDA. Compliance Program Guidance Manual 7346.832 Pre-Approval Inspections/ Investigations, 1990. Revised 1994.
2. FDA. Guide to Inspection of Pharmaceutical Quality Control Laboratories, 1993.
3. FDA. General Principal of Software Validation: Final Guidance for Industry and FDA Staff, 2002.
4. FDA. Industry Guide: Computerized Systems Used in Clinical Trials, 2007.

5. FDA. Electronic records; electronic signatures; final rule. Code of Federal Regulations, Title 21. Food and Drugs, Part 11. Federal Register, 62(54):13429–13466.
6. Pharmaceutical Inspection Convention, January 2002. Good practices for computerized systems in regulated 'GxP' environments. 2003.
7. GAMP. Guide for Validation of Automated Systems in Pharmaceutical Manufacture. Version 3. March 1998. [Version 4, December 2001].
8. Huber L. Validation of Computerized Analytical And Networked Systems. Englewood, CO: Interpharm Press, 2002.
9. PDA (Parenteral Drug Association, USA). Validation and Qualification of Computerized Laboratory Data Acquisition Systems (LDAS), 2000. Technical Paper 31.
10. FDA. Final Report on Its "21st Century" Initiative on the Regulation of Pharmaceutical Manufacturing. Available at: http://www.fda.gov/bbs/topics/news/2004/NEWOI120.html.
11. Huber L. Risk based validation of commercial off-the shelf (COTS) computer systems. Journal of Validation Technology 2005; 11(3), 184–204.
12. Huber L. Using macros & spreadsheets in a regulated environment, part of the macro & spreadsheet quality package. Labcompliance, May 2002. Available at: www.labcompliance.com/books/macros.
13. Crosson E, Campbell MW, Noonan T. Networking management in an FDA regulated environment. PDA Journal Nov–Dec 1999; 53(6):280–287.
14. Huber L. Using networks in a regulated environment, part of the network quality package. Labcompliance, April 2001. Available at: www.labcompliance.com/books/network-quality.htm.
15. Huber L. 21 CFR Part 11: Past, Present and Future, Journal of GxP Compliance, 2007; 12(4):22–37.

11

Integral for Successful PAI: The Quality Assessment Program

Graham Bunn

GB Consulting, Berwyn, Pennsylvania, U.S.A.

INTRODUCTION

The probability of a pre-approval inspection (PAI) success is significantly decreased if documentation supporting the final product is not complete, accurate, and in compliance with regulatory expectations, current good manufacturing practices (cGMPs), and customer expectations. The result of product development is the regulatory submission supported by enormous amount of data and information at various locations within and outside the company. Additionally, the commercial facility is required to have an infrastructure of systems with a wealth of documentation supporting each product production.

The number of data-entry points, calculations, reports, decisions, and recommendations, etc., which contribute to the development and also production of a product, is immeasurable. Contributions originate from multiple sources and converge in documents supporting systems. Relate this to a series of three-dimensional jigsaws carefully interlocked together with a defined interrelationship and dependency on each other. Removal of one piece of jigsaw e.g., relating to an inadequate report containing key decisions, appears to have little to no impact on the others. Depending on the order of the removal of other pieces relating to the critical data, executed batch record entries, etc. the degree of impact changes until there is an overall weakness in the structure leading to partial or total collapse. The failure of one system to provide essential infrastructure can result in immense problems for the host of the PAI.

The company's chief executive officer (CEO) constantly relies on all employees to meet, and often exceed, their responsibilities. Contributions are made to the overall documentation from operators recording the pH of a solution prior to filling, calculation of label accountability, recording analytical test results, line clearance confirmation, through decisions for product shelf life, investigation conclusions, and, ultimately, product disposition.

This chapter will examine the establishment of a program designed to monitor the compliance level of operations and leading to preparation for the PAI.

A paradigm shift from traditional internal audit program to quality assessment program (QAP) has been proposed and designed to be proactive (1). This program aligns with the issuance of the Food and Drug Administration (FDA) quality system inspection techniques (QSIT) (2) to examine documentation showing that management has established and consistently follows quality procedures.

It is time to refocus from the traditional audit of identifying what went wrong to being proactive and confirming system compliance. Operation of the systems can then be monitored to confirm that acceptable standards are being met and maintained.

Key support of the QAP to the PAI success relies upon

- absolute and complete support for the QAP from the CEO and direct reports,
- proactive and not reactive operations,
- customer and quality focus,
- actions prepared, executed, and implemented with "military precision,"
- striving for flawless operations,
- assigned and clearly defined responsibilities with accountability for meeting objectives and time lines,
- adequately managed changes with further explanations/justifications where necessary,
- addressed deviations and investigations,
- comprehensive corrective and preventative actions (CAPA) with appropriate time lines,
- coordination and clear communications (interdepartment, interdivision and intersite), and
- ensuring that any previous regulatory inspection compliance gaps commitments are adequately addressed or suitably justified for not completing the change.

Inputs to the product development program include

- approximately 750 million dollars for fundamental development (higher for more complex, biological products, specialized delivery systems),

- resources, including personnel, facilities, running costs, taxes, salaries, etc.,
- occupation of position in the development program,
- countless work hours, and
- enormous amount of data/information.

The cost of a PAI failure includes

- loss of confidence by the general public and stockholders,
- demotivation of company personnel,
- withholding of current, pending, or future applications until compliance status is remediated,
- reputation loss with the FDA, and
- incalculable financial loss from market sales.

Ask your CEO the question "Is PAI failure from a manageable/predictable situation acceptable?"

The answer is definitely no. The CEO should then be asked what specific influence is personally being applied to ensure that the PAI has a high degree of success. This sense of responsibility is the company's commitment to meeting customer/regulatory expectations.

Everyone contributing to each product production has an individual responsibility to ensure that they operate in compliance with cGMPs. The compliance function has the responsibility to confirm that company procedural requirements and cGMPs are being met.

The ultimate question is, "Are you in compliance if you have to prepare for a pre-approval inspection?"

This chapter will examine

- Essential background information for PAI preparation in relation to the QAP
- Coordination between research and development (R&D) and commercial site(s)
- QAP
  - Objectives
  - Resource requirements
  - Program design
  - Documentation

ESSENTIAL BACKGROUND INFORMATION

In order to develop a comprehensive program, it is essential to understand regulatory expectations and requirements. Personnel with key responsibilities and direct involvement must clearly understand the regulatory process of the PAIs.

The 1978 preamble to the cGMP published in the *Federal Register* (3) contains a specific reference to pre–new drug approval application (NDA) requirements:

> The Commissioner finds that, as stated in 211.1, these cGMP regulations apply to the preparation of any drug product for administration to humans or animals, including those still in investigational stages. It is appropriate that the process by which a drug product is manufactured in the development phase be well documented and controlled in order to assure the reproducibility of the product for further testing and for ultimate commercial production.

Note the selected words ''well documented and controlled'' and ''reproducibility.'' These terms are fundamental and core to cGMP regulatory expectations. They are independent of the stage of product (clinical trial or commercial).

The FDA also issued Guidelines on the Preparation of Investigational New Drug Products (Human and Animal) in March 1991. The guideline says that it represents the agency's current position on the requirements of the cGMP regulations. If a person chooses to depart from the practices and procedures set forth in the guideline, that person may wish to discuss the matter further with the agency to prevent an expenditure of money and effort on activities that may later be determined to be unacceptable by FDA. The guideline continues to clarify that cGMPs do not apply to initial R&D involving the drug's chemistry and toxicology conducted in laboratory animals (preclinical studies). It does, however, note that it was important to process materials under controlled conditions and to maintain adequate records.

Therefore, there is no doubt that the cGMPs in 21 CFR 210 and 211 apply to clinical supplies.

The current expectations of most FDA-regulated industries such as biotechnology and pharmaceuticals are no different than the following requirements, although the regulations do not specifically mandate the installation of an internal auditing program.

There is no question that the entire process of producing clinical supplies is covered by regulatory cGMP requirements. The definition of ''drug product'' in 21 CFR 210.3(b) (4) includes reference to placebo and, hence, clinical trial supplies. Data and information cumulated in the regulatory submission are reflected in the requirements at the commercial facility e.g., analytical methods and manufacturing processes (master batch records).

The Code of Federal Regulations (CFR) for medical devices, Quality System Regulations: 21 CFR 820, requires the following:

- Procedures for quality audits to be set up
- Such audits to be conducted to assure that the quality system is in compliance with the established quality system requirements and to determine the effectiveness of the quality system

- Quality audits to be conducted by individuals who do not have direct responsibility for the matters being audited
- Corrective action(s), including a reaudit of deficient matters, to be taken when necessary
- A report of the results of each quality audit, and reaudit(s) where taken, to be made
- Such reports to be reviewed by management having responsibility for the matters audited
- The dates and results of quality audits and reaudits to be documented

The key points of Good Laboratory Practice for Nonclinical Laboratory Studies (21CFR58) are as follows:

- Adequate inspection of each study with signed inspection records (defined content)
- Immediate reporting of problems to study director and management
- Periodic status reports (problems and corrective actions)
- Quality unit responsibilities defined in writing, with records/certification available for FDA inspection

This information should be appropriately incorporated into the QAP to ensure that the key points and related items are adequately covered.

Personnel responsible for the site QAP program need to be trained in regulatory expectations of the PAI, and the training includes the following program.

The FDA Compliance Program Guidance Manual (Program 7346.832) was revised in September 2003, and provides guidance for all aspects of the inspection, including objectives and responsibilities of the Center for Drug Evaluation and Research (CDER) and the district.

The following items may also be included in the responsibility of CDER but are listed here specifically for the district:

- Manufacture of biobatch: determination of the establishment's compliance with cGMP requirements, including a data audit of the specific batches upon which the application is based, e.g., pivotal clinical, bioavailability, bioequivalence, and stability
- Manufacture of drug substance: determination of cGMP compliance
- Manufacture of excipients: cGMP inspections usually performed at the request of CDER
- Raw materials: establishment inspection of drug substance and review of raw material data to determine cGMP compliance
- Raw materials (tests, methods, specifications): audit of data submitted in the application
- Composition and formulation of finished dosage form: audit of data submitted in the application

- Container/closure system(s): audit of the data submitted in the application
- Labeling and packaging controls: determination of the cGMP compliance and audit of the data submitted in the application
- Laboratory support of methods validation: upon CDER request, field laboratory analysts will conduct laboratory validation of the analytical methods proposed by the applicant
- Product controls: establishment inspection to determine compliance with cGMP requirements, and review and audit of the data submitted in the application
- Product tests, methods, and specifications: audit of the data submitted in the application
- Product stability: establishment inspection to determine compliance with cGMP requirements, and review and audit of the data submitted in the application
- Comparison of the relevant pre-approval batch(es) and proposed commercial production batches: comparison of process to manufacture the pre-approval batches with the actual process used to manufacture the validation batches
- Facilities, personnel, equipment qualification: review of information and establishment inspection to determine compliance with cGMP requirements
- Equipment specification(s): audit of the data submitted in the application
- Packaging and labeling: review of the cGMP controls and inspection of the establishment to determine compliance with cGMP requirements
- Process validation: inspection of the establishment to determine compliance with cGMP requirements and adherence to application requirements
- Reprocessing: inspection of the establishment to determine compliance with cGMP requirements and to audit the data in the application, including the validation data
- Ancillary facilities: facility review and inspection of ancillary facilities (contract testing laboratories and contract packages and labelers) at the request of CDER or as a result of the district management's judgment

A comprehensive, system-based assessment program will cover the above items. This list is not intended as an all-encompassing checklist for the QAP and, in turn, pre–PAI preparation. These are key input to the overall ongoing QAP.

QUALITY SYSTEM INSPECTION TECHNIQUE

The FDA's Office of Regulatory Affairs and the Center for Devices and Radiological Health issued the Guide to Inspection of Quality Systems in August 1999 and changed the approach to inspections for the medical device industry (2). Inspections historically involved reviewing lists, linking to documents, and procedures to determine compliance with regulations and expectations. This approach was referred to as a "bottom-up" approach because of the review of output from systems. The QSIT is a "top-down" approach to inspecting subsystems that are identified as the key elements of a company's quality system to efficiently and effectively evaluate that quality system. The following four key systems are covered:

- Management control
- CAPA with satellites medical device reporting, corrections and removals, and medical device tracking
- Design controls
- Production and process controls with satellite sterilization process controls

The overall process for each system involves determining if the requirements are defined and documented by reviewing the procedures and policies. Using sampling tables, the raw data in records are reviewed to determine if the requirements are being implemented and adequate. The Guide to Inspection of Quality Systems provides investigational objectives and a decision flow chart with supporting narrative for each of the systems.

The FDA's alternative inspection approach was outlined for drug manufacturing inspections in Compliance Program Guidance Manual Program 7356.002 with an implementation date of February 1, 2002. It applies to all drug manufacturing operations. The goal of the program's activities was to "minimize consumer's exposure to adulterated drug products." The audit covers two or more systems, which always includes the quality system.

The following systems are covered:

- Quality: overall compliance with cGMPs, internal procedures, and specifications. This includes the quality control unit and all of its review and approval duties (change control, batch release, reports etc.). See 21 CFR 211 subparts B, E, F, G, I, J, and K.
- Facilities and equipment: measures and activities, which provide appropriate physical environment and resources, buildings, and facilities, including maintenance, equipment qualifications, calibration and preventative maintenance, cleaning, and validation of cleaning of processes. Process performance qualification will be evaluated as part of the inspection of the overall process validation, which is done within the system where the process is employed. See 21 CFR 211 subparts B, C, D, and J.
- Materials: measures and activities to control finished products, components, containers, and closures. It also includes validation of computerized inventory control process, drug storage, distribution controls, and records. See 21 CFR 211 subparts B, E, H, and J.
- Production: measures and activities to control manufacture of drugs and drug products, including batch compounding, dosage for production,

in-process sampling and testing, and process validation. Establishing, following, and documenting performance of approved manufacturing procedures are also covered. See 21 CFR 211 subparts B, F, and J.

- Packaging and labeling: measure and activities controlling the packaging and labeling of drugs and drug products. It includes written procedures, label examination, label storage and issuance, packaging and labeling operations controls, and validation of these operations. See 21 CFR 211 subparts B, G, and J.
- Laboratory control: measures and activities related to laboratory procedures, testing, analytical methods development and validation or verification, and the stability program. See 21 CFR 211 subparts B, I, J, and K.

Organization and personnel with qualifications and training are evaluated within each of the systems. Inspection of contract companies is included within the system for which the product or service is contracted as well as their other quality systems.

Additional guidelines and supporting documents are available on the FDA Web site (www.FDA.gov) and are listed at the end of this chapter.

COORDINATION BETWEEN R&D AND COMMERCIAL SITE(S)

The R&D sites submitting data/information supporting the application are open to inspection. The key considerations are as follows:

- Acknowledgment by senior management that the R&D facility/facilities has/have the potential for an inspection
- Preparing a comprehensive long-term preparation plan to ensure PAI readiness with the commercial facility
- Determining the resource requirements to ensure that all operations are continuously evaluated against predetermined standards/criteria
- Defining key milestones to ensure that documentation supporting each product is accurate and complete at that point in time according to procedural requirements
- Defining individual and overall responsibilities for ensuring readiness for the inspection

Drug development from the time of investigation new drug (IND) filing to the that of new drug application (NDA) approval takes approximately 4.5 to 9 years and is influenced by many factors, e.g., number of clinical studies and patient recruitment time. The volume of documents generated during development is vast and often includes multiple sites producing the clinical supplies and ultimately the commercial product, in multiple locations around the world. The development report (4) covers key aspects of the product, which must be accurate, complete, and verified by the R&D QAP function. It is essential that

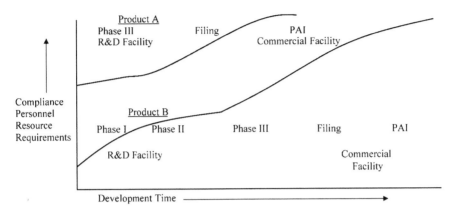

Figure 1 Graph of compliance personnel resources versus development time.

there be open communication between the assessors responsible for R&D (between the multiple sites) and those responsible for the commercial facilities. This communication can be effectively achieved by the exchange of team members between the two QAP team functions.

The process is further complicated by inevitable changes to the documentation supporting the drug product as it is developed over several years. For example, the formulation (captured in the master batch records) and key process steps undergo several changes during development of the final product and are managed through process change control.

The graph above (Fig. 1) illustrates the development of products A and B, which could have data primarily generated at the same site or multiple sites. In this example, they are being transferred to the same commercial site. The two PAIs may in practice be separated by several months but additional resources may be warranted to manage this temporary situation.

The graph illustrates the increase in compliance resources as the product progresses through development. It is important to remember that primary discussion in this chapter is in relation to a single product but in practice, companies have several products at various stages of development. Products can be removed at any time during development because of clinical trial results, formulation technology problems, or other business factors.

There needs to be a well-designed technology transfer program for the product to be ultimately manufactured and tested at the commercial site. This program also includes the entire infrastructure supporting the commercial manufacturing, e.g., stability, chemical, biological, and microbiological testing (where applicable).

The drug development process involves a complex process over an extended period. There is no substitute for the documentation and data generated during this time, and while some may be constructed from supporting data afterward, it is not always possible. Valuable time can be lost and ultimately information/data may not be available.

There must be a clearly defined QAP administered by a quality function, which begins from the time data and information is generated to support each IND. The program should be integrated into project management for not only the R&D portion but also for the transfer to commercial production. This project can be further complicated by data/information being generated by multiple R&D and may be transferred to more than one commercial site (manufacturing, packaging/labeling).

Another dimension is added when contractors are used for portions of manufacturing, packaging, and testing. The customer maintains specific documents (e.g., batch records), but it is essential that access to contractor supporting information/data be provided.

QUALITY ASSESSMENT PROGRAM

Objectives

Preparation for a PAI is a continuous process as new products enter the development program and are ultimately transferred from clinical production to the commercial facility. An assessment program needs to maintain the compliance status of the facilities contributing to the submission and those where commercial production takes place. The absence of an ongoing effective program or focusing assessments immediately prior to the PAIs does little to assure confidence of a successful inspection.

The key to the success of the QAP is senior management support, including local facility senior management and corporate executive management. The program cannot be implemented or be effective without direct management involvement and support. The QAP standard operating procedures (SOPs) must define the responsibilities of the individuals involved and the interface to reporting compliance gaps to the various management levels.

Resource Requirements

The assessors must have the necessary education, skill set (general and specialized, according to their responsibilities, e.g., vaccines, sterile, dosage forms, biotechnology, and interpersonnel communication), and experience to support a QAP. There should be a comprehensive training program to orientate new assessors to the company standards and requirements. Ongoing education integrated with continuous evaluation of regulatory expectations and industry standards are essential in ensuring that assessors remain current with GMP requirements.

There needs to be adequate personnel to support the assessment program requirements with sufficient administrative functions.

Program Design

Coordination of compliance monitoring can include multiple commercial sites and often one or more contractors. Outsourcing to contractors has become more critical to product development and commercial supplies can involve the entire process from active pharmaceutical ingredients to shipping final product. Variations of this complete service can involve different contractors in different locations, and sometimes in other countries.

The QAP must incorporate the entire manufacturing, packaging, and testing requirements with all parties. Coordination of these activities is paramount because regulatory observations at any one of the facilities could jeopardize success of the PAI. Corporate compliance should ensure that there is adequate communication between site compliance functions when product production crosses multiple divisions, including countries. Compliance responsibilities for contractors may be managed by a separate group but must be integrated into the overall site QAP in preparation for commercialization of the product. This is illustrated in the following diagram (Fig. 2).

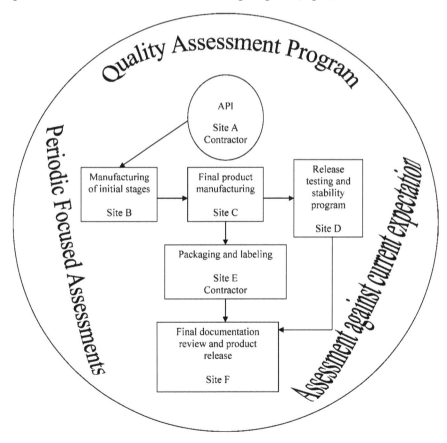

Figure 2 Quality assessment program.

The product being introduced into the commercial facility may require special requirements, e.g., 5°C storage or testing using new equipment. The QAP must incorporate these specific requirements and ensure that they are accommodated in the overall compliance systems without interference with existing compliance status. Project planning is critical to product introduction to manage resources, timelines, and to ensure that all required aspects are adequately covered.

The QAP is a team function and therefore it is essential that the conventional misnomer of auditor and auditee be dismissed. Assessor and host are more appropriate and align with the proactive orientation of the QAP. This is a collaborative effort with a single objective to operate in a state of compliance. There are no rewards for identifying compliance gaps, only for ultimate product approval following PAI.

Conventional Audit

- The auditor makes observations based on following product process flow.
- The auditee responds to observations.
- The auditor evaluates responses and timelines and revisions may be performed by the auditee to meet expectations.
- The auditor follows up to ensure corrective actions are completed.

Alternative Approach to PAI Preparation

- Coordination between assessment team and host(s)
- Collaboration between host(s) and subject matter experts to perform system-based assessment
- Identification of compliance gaps
- Investigation of gaps including root causes and determination of compliance risk factors
- Agreement of responses including time lines between leader assessor and host(s)
- Incorporation of CAPA into monitoring program
- Verification of adequate/complete closure of actions according to time lines

The QAP must be defined in an SOP and include the following key points:

- Schedule covering all the systems
  - Some systems/areas may be
    - assessed more frequently depending on factors including compliance gaps, regulatory risk, priorities, etc.;
    - issued to senior management (usually annually) and revised periodically with justification for any delays and notification updates to management; and
    - refocused as necessary according to identified compliance gaps.

- Lead assessor responsibilities
  - Collection of all available current information (drug development report, regulatory submission, cGMP expectations, previous compliance gaps)
  - Creation of plan defining scope, dates, and any exclusions/limits
  - Liaison with the host for scheduling and logistics
  - Establishment of a multidisciplinary team
  - Coordination with other team members to ensure that any specific expertise/skills are represented during the assessment
  - Coordination on site to ensure plan objectives are met
  - Coordination of the assessment report drafting with input from team members
  - Issuance of the report, including compliance gaps to management and the host with a defined date for responses
  - Evaluation of the responses with corresponding completion dates and liaison with the host to resolve any differences
  - Issuance of the final report to management
  - Tracking of responses according to completion dates and periodically reporting status to the management

Management is procedurally responsible for responding ultimately to the CEO for any overdue commitments. Response closures and metrics need to be a standing agenda item on the senior management meeting, and the responsible directors are required to justify overdue responses.

- Team membership
  - Team assessments are effective when including personnel with different expertise. The lead assessor is responsible for coordinating each team member's specific responsibilities, logistics, and input.
  - It is imperative that the personnel responsible and those participating in the assessment program are adequately informed of all regulatory expectations and closely familiar with the FDA Compliance Program Guidance Manual on Pre-approval Inspections/Investigations (Program 7346.832).
  - It is essential that the personnel remain current with regulatory expectations through all information sources, which include
    - FDA-issued guidelines
    - Seminars
    - Podium presentation
    - Warning letters
    - FD 483s
    - FDA Web site
    - Recalls

- Consent decrees
- Compliance influenced by advances in science and technology

Documentation

- The SOP defining the assessment report will cover
  - Executive summary content
  - Key compliance gaps format
  - Personnel to which report is issued
  - Time lines for report issuance after assessment and time for host to respond
  - Final report issuance
  - Control of report copies

The assessment SOP should be linked to the CAPA system. Response initiated because of the QAP must be monitored to ensure adequate and timely completion.

$$\text{Assess} \rightarrow \text{Identify} \rightarrow \text{Collate} \rightarrow \text{Evaluate} \rightarrow \text{Respond} \rightarrow \text{Verify}$$

Assess systems against standards to *identify* compliance gaps and then *collate* results. *Evaluate* the results in order to *respond* and then *verify* that the responses meet commitments within the defined timeline.

A SOP should define the documentation evidence of the QAP that is produced, if requested by regulatory authorities during the PAI.

METRICS

The compliance assessments for the PAI performed through the QAP provide an incomplete compliance status and inadequate support for the PAI program if the data are not fully utilized. It is essential that the responses are captured in the CAPA system and tracked to ensure adequate completion by the defined date. Metrics for monitoring performance of the QAP system and analysis of the compliance gaps need to be routinely reported to senior management. The analysis may include number of compliance gaps in defined systems assigned a severity according to agreed criteria. Included with the metrics are recommendations for management concurrence, which may include resources of both personnel and monetary as necessary.

A risk management standard from the joint Australian and New Zealand Standards on Risk Management (AS/NZS 4360) (5) proposed a matrix of probability and severity to qualitatively assess risks and provide a means of prioritizing them for control.

This approach can be used for presentation of compliance gaps and enables the management to effectively focus on the areas requiring immediate actions (shaded in the example on page 257). For example, after numbering the compliance gaps they are collated according to the probability and ranked according to

Probability Vs. Severity Matrix (AS/NZS 4360)

		Insignificant	Minor	Moderate	Major	Catastrophic
Probability	*Almost Certain*	Significant	Significant	High	High Gap #3	High
	Likely	Medium	Significant	Significant	High Gap #1	High
	Moderate	Low Gap #2	Medium	Significant	High	High →
	Unlikely	Low	Low	Medium	Significant	High
	Rare	Low Gap #6	Low Gap #4	Medium	Significant Gap #5	Significant
		Insignificant	*Minor*	*Moderate*	*Major*	*Catastrophic*
		Severity				

severity. The corresponding numbers of the gaps are entered in the table as seen above for numbers 1 through 6. The scattering of the numbers provides an overall compliance "picture" and a tool to determine changes between each assessment performed.

Models that consider the level of compliance and the effectiveness have been proposed (6). Proactive compliance management requires the use of compliance effectiveness profiling (7). The management can determine from the differences between actual and target compliance operating levels the degree of risk or if they are "over insured" (actual is greater than target). Location of the company operating level in one of the three areas in the figure below (Fig. 3) informs management if the infrastructure is optimum, cost-effective, and capable of more or over costly for what is being delivered. The growth area is seen up to the peak of the effectiveness and after this is decay.

The cost of the level of compliance can be seen in the figure below (Fig. 4). The cost is usually the dollar amount but can also interestingly be interpreted as the results of regulatory compliance, i.e., FD 483s or warning letters.

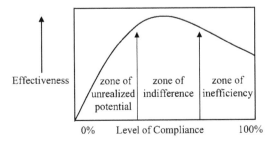

Figure 3 Effectiveness of compliance versus level of compliance.

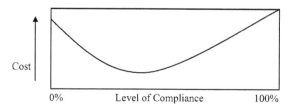

Figure 4   Cost versus level of compliance.

CONCLUSION

It may be concluded from the above that there is a minimum cost for a level of compliance that has an acceptable risk (regulatory action, not patient product). Beyond a certain level, the return on investment cost may outweigh the level of compliance that in practice may not be achievable. It is doubted that there is a regulated industry that is in a constant state of compliance with all regulatory expectations. This doubt is because the industries are complex and in a constant state of change. Hence change can be managed and regulatory risk minimized with an effective compliance program.

Those parts of the system, which failed and resulted in the compliance gap, must be identified. The compliance gap is a result or a symptom of an underlying cause that needs to be traced back to the root cause. Only then can the analysis of the compliance gap be performed and suitable corrective action plan be composed for implementation. Nonsystematic errors should also be identified, i.e., those that are not a direct contribution (independent) of the system. These may include variability in the input from people, which is not a direct system failure of procedure or documentation. These errors should be equally investigated and appropriate action taken. Corrective action for the failure of a person to complete a task or follow a procedure cannot always be retraining. If the initial training was ineffective, there may be other causes. Hence, no amount of training will effect change of personnel behavior.

Global management can achieve collaboration between company sites in preparation for the PAIs. Sharing of compliance gaps ensures that resolutions are initiated across sites and departments.

While a comprehensive and well-executed QAP serves as a sound foundation for the PAI, there is also additional benefit from an overall independent review. This assessment should be performed at a time suitable to enable sufficient time to perform any corrective actions or implement a contingency plan. The assessment can also include a mock PAI specifically designed to train personnel who interact with the FDA during the actual PAI.

Compliance with cGMPs and regulatory expectations is the responsibility of everyone employed in the regulated industries. It starts from the moment data/ information is captured supporting a regulatory submission, including all clinical trial products. There is no finite end to compliance requirements while product is still in the market. Regulatory requirements are always changing and regulatory compliance is based on being a leader, not a follower. The QAP confirms and verifies the current compliance status. The ultimate goal is to be in a constant state of compliance and, therefore, pre-PAI assessment should be a confirmation, not a preparation.

Meeting minimum standards and expectations should be replaced by meeting company standards, which are based on customer/regulatory expectations. Leading and pioneering standards do not necessarily result in setting unobtainable, or even unnecessary, goals. The achievable level of compliance needs to be value added, the way to conduct regulated business, and not a reflex to regulatory compliance gaps.

REFERENCES

1. Bunn G. The internal audit program. In: Medina C, ed. Compliance Handbook for Pharmaceuticals, Medical Devices and Biologicals. New York: Marcel Dekker, 2004:429–426.
2. FDA. Quality System Inspection Technique, August 1999.
3. FDA. Preamble to Current Good Manufacturing Practice in Manufacture, Processing, Packing, or Holding [docket no. 75N-0339] March 28, 1979.
4. Freeman S. Change management: a far-reaching, comprehensive, and integrated system. In: Medina C, ed. Compliance Handbook for Pharmaceuticals, Medical Devices and Biologicals. New York: Marcel Dekker, 2004:315–345.
5. Australian and New Zealand Joint Standard on Risk Management (AS/NZS 4360), Appendix A:20, 1995.
6. Juran JM, Gryna F. Quality Planning and Analysis. 3rd ed. New York: McGraw-Hill, 1993.
7. Dean D, Bruttin F. Risk and the economics of regulatory compliance. PDA J Pharm Sci 2000; 54(3):253–263.

FURTHER READING

- 7346.832 CPGMP Pre-approval Inspections/Investigations.
- 7356.002 CPGMP Drug Manufacturing Inspections.
- 40832 NDA Pre-approval Inspections (Bacterial & Allergenic Products).
- 42832 NDA Pre-approval Inspections (biological products including urokinase).
- 42R806 Foreign Inspections (NDA, BLA, where CBER is the lead center).
- 46832 NDA Pre-approval Inspections/Methods Validation.
- 46832B NDA Forensic Sample Collection/Analysis.
- 46R806 NDA Foreign Inspections.
- 52832 ANDA Pre-approval Inspections/Methods Validation.
- 52832B ANDA Forensic Sample Collection/Analysis.
- 52832C ANDA Biotest Sample Collection/Analysis.
- 52R806 ANDA Foreign Inspections.

12

All Dressed Up but No Approval to Go: The Consequence of Failing an FDA Pre-Approval Inspection

Alan Minsk and David Hoffman[a]

Arnall Golden Gregory LLP, Atlanta, Georgia, U.S.A.

The Federal Food, Drug, and Cosmetic Act (FDCA) provides that the Food and Drug Administration (FDA) may only approve a new drug application (NDA), an abbreviated NDA, or a biologics license application if the methods used and the facilities and controls for the manufacture, processing, packing, and testing of the drug or biologic are found adequate to ensure the drug's identity, strength, quality, and purity. An applicant is required to provide information to the agency, demonstrating the method of analysis and details specifying how the company plans to manufacture the proposed product. The reviewing division will evaluate data and information submitted by the applicant and establish specifications for the manufacture and control of the resulting drug product on the basis of the submitted data.[b]

---

[a]Alan Minsk is a partner and chair of the Food and Drug Practice Team at Arnall Golden Gregory LLP, based in Atlanta, Georgia. David Hoffman is an associate with the Food and Drug Practice Team. Messrs. Minsk and Hoffman can be reached at alan.minsk@agg.com and david.hoffman@agg. com, respectively.

[b]Although the review division of the agency is responsible for determining whether a new drug application is acceptable, the compliance division or appropriate FDA District office will conduct the actual pre-approval inspection and report to the review division its findings.

As a condition for approving an application, the FDA must ensure that the applicant will comply with the current good manufacturing practices (cGMPs), verify the authenticity and accuracy contained within such applications, and determine if other information may affect the applicant's ability to manufacture the product in compliance with cGMPs. Such a determination is often made through a pre-approval inspection. Because the validity of the applicant's manufacturing facilities is critical to determine whether the agency may approve an application, the inspection must be done within the timelines for review established under the FDCA and the regulations.[c] Such an inspection can extend to a contract manufacturer and those firms involved in the manufacturing process.

The FDA inspections, including pre-approval inspections, are to occur at a reasonable time, with reasonable limits, and in a reasonable manner.[d] The FDCA requires that inspections be conducted every two years, unless there is a reason to inspect earlier. The inspection "shall extend to all things therein (including records, files, papers, processes, controls, and facilities)" bearing on whether the products are manufactured in a compliant manner (e.g., not adulterated, not misbranded).[d] The FDA inspections can cover the following: (1) premises and all pertinent equipment; (2) finished and unfinished materials; (3) containers; and (4) labeling within the establishment or vehicle in which products are manufactured, processed, packed, held, or transported. The agency can also take samples of regulated products and labeling.

The division of the FDA reviewing the application may request a pre-approval inspection of the sponsor's manufacturing facilities and clinical trial sites. The decision to submit an application begins the review process and, when needed, initiates a request for a pre-approval inspection. During such inspections, FDA investigators audit manufacturing-related statements and commitments made in the application against the sponsor's manufacturing practices. More specifically, the FDA conducts inspections to

- verify the accuracy and completeness of the manufacturing-related information submitted in the application;
- evaluate the manufacturing controls for the pre-approval batches upon which information provided in the application is based;
- evaluate the manufacturer's compliance with GMPs and manufacturing-related commitments made in the application; and
- collect a variety of product samples for analysis by the FDA laboratories. These samples may be subjected to several analyses, including methods validation, methods verification, and forensic screening for substitution.

---

[c]For example, under the Prescription Drug User Fee Act, the agency must determine whether a new drug application or abbreviated new drug application is approved, approvable, or not approved within 180 days. See 21 CFR § 314.100.

[d]21 USC § 374(a)(1).

According to an FDA policy, product-specific pre-approval inspections generally are conducted for products (1) that are new chemical or molecular entities, (2) that have narrow therapeutic ranges, (3) that represent the first approval for the applicant, or (4) that are sponsored by a company with a history of cGMP problems or a company that has not been the subject of a cGMP inspection for a considerable period of time.[e]

The results of the pre-approval inspection may affect the agency's final approval decision. If a pre-approval inspection discovers significant cGMP problems or other issues, the reviewing division may withhold approval until the issues are addressed and corrected. The division's response to such deficiencies will likely depend on several factors, including the nature of the problem, the compliance history of the facility, the prognosis for the problem's correction, and the potential effect of the problem on the safety and efficacy of the drug.

The FDA has a number of enforcement options if it believes that a manufacturing site is in noncompliance with the law. Such options include, but are not limited to, the FDA issuing an FD-483, which is a list of observations of possible violations, sending a warning letter, issuing import alerts or detentions, or seeking court relief, such as product seizure or injunction.[f] For a pre-approval inspection, failure to meet the FDA's requirements may lead to a delay in an application's approval.

Such a delay, although not an outright rejection, can have significant financial ramifications, which will negatively affect the entire company. In addition, of course, patients who need the treatment will also be affected by such a delay.

The FDA notes in its warning letters that failure to comply with the FDCA and the FDA's implementing regulations can lead to product approval delays. For example, in a warning letter to Abraxis Bioscience, Los Angeles, California, U.S.A., dated December 18, 2006, about its manufacturing plant in Melrose Park, Illinois, the FDA noted:

> Until FDA can confirm correction of the deficiencies observed during the most recent inspection, this office can recommend disapproval of any new applications listing this site as a manufacturer of drugs.

Similarly, the agency wrote in a warning letter to Actavis Totowa, LLC, Little Falls, New Jersey, U.S.A., dated February 1, 2007:

---

[e]More specific guidance on CDER's pre-approval inspection program is available from CDER's Compliance Program Guide 7346.832. The CPG is available on FDA's Web site: http://www.fda.gov/ora/cpgm/default.htm.

[f]If a company disagreed with the results of a pre-approval inspection, the company could seek to appeal the decision. If the issue was a technical one, a company could use CDER's January 2006 "Formal Dispute Resolution for Scientific and Technical Issues Related to Pharmaceutical cGMPs." A company may also use the FDA's citizen petition procedures to seek a review of an adverse decision. See e.g., 21 CFR § 10.33; § 10.45.

FDA may withhold approval of pending new drug applications listing your facility as a manufacturer until the above violations are corrected.

The FDA language is not a mere threat. The delayed approval of Schering-Plough's Clarinex® antihistamine, the successor to Claritin®, for nearly a year was, in part, due to manufacturing plant deficiencies. Similarly, the authors are aware of a situation where a client received an approvable letter, whereby the FDA said that the application was approvable, but for a final inspection of the contract manufacturer. However, the contract manufacturer did not pass its inspection and, therefore, the agency was prepared to delay approval of the application. The CEO of the drug company client whose application was deemed approvable contacted the authors to emphasize the importance of obtaining the FDA approval of the drug product. Only after successful negotiations with the FDA did the agency ultimately approve the drug company's application, despite the contract manufacturer's cGMP deficiencies.

Companies understandably fear warning letters. However, senior management sometimes does not take as seriously as it should the consequence of a poor inspection, delaying approval or even leading to the rejection of an application. The financial consequences are significant, such as no immediate sales revenues, individuals hired to sell an unavailable product, stock price drop, and disheartened employees. This loss doesn't take into account the cost of hiring consultants and lawyers. The consequences of a poor inspection have long, ripple effects that can hurt a company's growth, if not crush the firm, if it is the one or only product.

Many people take painstaking efforts to maintain their car, their home, their garden, or other property interest. Such proactive attention to detail is imperative for ensuring a quality manufacturing facility. Training, employing top-notch employees, instituting a compliance first policy, and monitoring and auditing are only some activities that can be taken to maximize a good pre-approval inspection and minimize unpleasant surprises.

Index

9 780367 452711